LTE FOR PUBLIC SAFETY

LTE FOR PUBLIC SAFETY

Rainer Liebhart, Devaki Chandramouli, Curt Wong and Jürgen Merkel

Nokia, Networks Division

Library of Congress Cataloging-in-Publication Data applied for

A catalogue record for this book is available from the British Library.

ISBN: 9781118829868

Set in 10/12pt, TimesLTStd by Laserwords Private Limited, Chennai, India

1   2015

Table of Content

Foreword

Foreword by Dr. Hossein Moiin

People love Long-Term Evolution (LTE). At 280 million LTE subscribers globally in October 2014, LTE has been adopted faster than any previous generation of mobile technology. The time to reach one billion subscribers is expected to be 7 years, as compared to 11 years for 3rd Generation (3G) and 12 years for Global System for Mobile Communications (GSM). LTE also has the biggest and fastest growing device ecosystem. With 331 commercially launched LTE networks in 112 countries today and more than 600 operator commitments, it has become the technology of choice for operators. I believe that LTE will outgrow other technologies to become the unrivaled fabric of mobile broadband, thanks to its feature richness, wide spectrum availability, and economy of scale.

As a technology, LTE has many unique characteristics that make it suitable for extension to services beyond basic mobile broadband. LTE is not only built upon improved radio principles with peak data rates beyond 150 Mbps, scalable bandwidth, and low latency but also has a much simplified Internet Protocol (IP)/Internet-based two-node architecture. As a result, LTE broadens the reach of mobile communications in the future by providing a platform for services such as LTE for Public Safety, LTE for TV Broadcast, LTE in Unlicensed and in UHF bands, LTE for M2M, LTE for Device-to-Device communication, and many other applications such as LTE for Connected Cars and for Airplanes.

Public Safety networks in particular benefit tremendously by using LTE as the base technology, which not only fulfills the specific communication needs of emergency services such as robustness and low latency but is also supported widely in commercial cellular systems already in use by people all over the world. Equally important is the fact that LTE for Public Safety standardization work benefits from the excellence in global standardization processes achieved by 3rd Generation Partnership Program (3GPP), as demonstrated in the superior craftsmanship of LTE that met user expectation with the first release and had the fastest and most stable standards.

Nokia Networks is actively driving the LTE for Public Safety standardization work in 3GPP and is the lead rapporteur in 3GPP SA/CT WGs for group communication, an essential feature for Public Safety. Dedicated efforts led by the authors of this book, Rainer Liebhart, Devaki Chandramouli, Curt Wong, and Jürgen Merkel, have been instrumental in elevating LTE for Public Safety, from a niche topic facing much resistance into the most supported system-level work in 3GPP Release 12.

I believe that LTE is the right technology for evolving Public Safety networks over mobile broadband, offering totally new and much more efficient ways for Public Safety personnel to communicate in the future. I do like to thank the authors for bringing out this book that will definitely help readers gain a detailed understanding of the technology behind LTE for Public Safety and the related aspects such as spectrum, architecture, features, interworking, and deployment scenarios.

Hossein Moiin
EVP, Technology & Innovation Leader
Nokia, Networks Division

Foreword by Mr. Andrew Thiessen

Public Safety's selection of the 3GPP LTE signals the beginning of a foundational movement by Public Safety away from the niche technology of the past – Land Mobile Radio (LMR) – to an advanced technology being embraced by the commercial marketplace. This move will allow Public Safety to reap the benefits of a market many times the size of the LMR market. Not the least of these benefits is access to a never-ending stream of technology refresh, something that has never been available to narrow-band voice systems.

The 3GPP community has shown significant support of Public Safety's decision to move to LTE. This is evident in the progress being made in closing the gaps between Public Safety's current mission critical voice requirements and the features provided by LTE. Specifically, the creation of Proximity Services (ProSe) within 3GPP will allow Public Safety user devices to communicate without infrastructure, which the Public Safety community in the United States considers fundamental to a mission-critical-capable technology. Similarly, the addition of Group Communications Systems Enablers for LTE (GCSE_LTE) will provide for efficient group communications, which is the predominant method by which the first responder community communicates. 3GPP made an additional commitment to the Public Safety community with the creation of a new working group chartered to work on applications and services, first among which will be mission critical push to talk (MCPTT).

So, for the first time, the global Public Safety community is coming together in a single standards development organization, around a single technology, to work collaboratively with commercial mobile network operators and equipment providers to move the world toward a communications environment that will more effectively support first responders as they carry out their vital mission to protect lives and property.

This book provides a detailed synopsis of recent achievements made in 3GPP to advance this important work; in particular, it describes the Public Safety specific features ProSe and GCSE_LTE. The authors are deeply involved in the work on LTE for Public Safety and are in a position to provide valuable firsthand insights on 3GPP and the relevant technical details.

Andrew Thiessen
Deputy Program Manager
U.S. Department of Commerce
Public Safety Communications Research Program

About the Authors

Rainer Liebhart has over 20 years of experience in the telecommunication industry. He held several positions within the former Siemens Fixed and Mobile Networks divisions and now in Nokia Networks division. He started his career as SW Engineer, worked later as standardization expert in 3rd Generation Partnership Project (3GPP) and European Telecommunications Standards Institute (ETSI) in the area of Internet Protocol Multimedia Subsystem (IMS), took over responsibilities as Worldwide Interoperability for Microwave Access (WiMAX) and Mobile Packet Core System Architect and is currently heading the Mobile Broadband Core Network standardization team in Nokia Networks. He is the Nokia Networks main delegate in 3GPP SA2 with the focus on Long-Term Evolution/System Architecture Evolution (LTE/SAE). He is (co-)author of over 50 patents in the telecommunication area. Rainer Liebhart holds an MS in Mathematics from the Ludwig-Maximilians University in Munich, Germany.

Devaki Chandramouli has over 14 years of experience in the telecommunication industry. She spent the early part of her career with Nortel Networks and is currently with Nokia, Networks division. At Nortel, her focus was on design and development of embedded software solutions for Code Division Multiple Access (CDMA) networks. Later, she represented Nortel in the WiMAX Forum and worked on WiMAX architecture and protocol development. At Nokia, her focus areas include architecture development within 5G research and standards development for 3GPP access, LTE/SAE-related topics. She is also an active participant and contributor to 3GPP standards. She has (co-)authored over 40 patents in wireless communications. She received her BE in Computer Science from Madras University (India) and MS in Computer Science from University of Texas at Arlington (United States).

Curt Wong has over 18 years of experience in the telecommunication industry. He started his career with Nokia Networks since 1995 and held several positions, including the area of research and development, interoperability testing, product management, and new technologies standardization. In recent years, his primary focus has been in the areas of wireless system-level developments, with emphasis on the infrastructure side of cellular networks. His current activities and involvement are on voice services over LTE, including IMS, emergency aspects, group communication over LTE, and interworking with legacy 2G and 3G networks. He is an active participant and contributor to 3GPP standards. He has a BS in electrical engineering from the University of Texas at Austin and an MS in telecommunications from Southern Methodist University, Dallas, TX.

Jürgen Merkel has over 20 years of experience in the telecommunication industry. He started his career in Alcatel where he held several positions in system design, product management, and business development and strategy – always with a tight linkage to standardization in ETSI Special Mobile Group (SMG) and 3GPP. In Nokia, he is a member of the Mobile Broadband standardization team, focusing on services and service requirements. He heads the Nokia Networks delegation in 3GPP SA1, and is an active contributor and driver for group communication aspects in 3GPP. Jürgen Merkel holds a MS in electronic engineering from the University of Stuttgart, Germany.

Preface

Long-Term Evolution (LTE) has turned out to be a huge success story not only technology wise but also commercially. At the time of writing this book, more than 300 LTE networks are deployed around the world. The number of available LTE user devices is close to 1900 (figures are taken from Global Mobile Suppliers Association (GSA)). LTE technology allows addressing a wide market, potentially all Global System for Mobile Communications (GSM) and Universal Mobile Telecommunications System (UMTS) network operators, and also Code Division Multiple Access (CDMA) operators and even operators providing converged fixed and mobile networks. LTE provides in-built support for interworking with GSM/EDGE Radio Access Network (GERAN), Universal Terrestrial Radio Access Network (UTRAN), CDMA, Wireless Local Area Network (WLAN), and fixed broadband access. With 3rd Generation Partnership Program (3GPP), LTE is backed by a strong standardization organization. Corrections and enhancements to the LTE system are provided rapidly and are based on consensus among all major device and infrastructure vendors. In addition, LTE can operate in a large variety of frequency bands. In the future even unlicensed bands might be supported. All this together provides strong confidence in the LTE technology, in the standardization process, and in a reasonable total cost of ownership for all players (operators as well as vendors). Thus, LTE was a natural choice as the future mobile broadband radio technology for Public Safety networks not only in particular markets such as the United States or Europe but around the whole globe. LTE has the potential to replace existing narrowband Land Mobile Radio (LMR) systems that are currently in use, providing a variety of new and sophisticated services beyond voice to Public Safety personnel.

The main intention of this book is to explain how LTE can be used as technology enabler for Public Safety networks. For that purpose, we focus on describing the new Public Safety related features that were standardized in 3GPP Release 12 on top of LTE. As we do not require that all readers are fully familiar with LTE and its basic concepts, we give an overview of this technology and some of the most important services available in LTE, providing also a justification why LTE was adopted for future Public Safety networks. The reader can find a more detailed description of the book's content in the introduction section.

The book is intended for a variety of readers such as students, network operators offering their LTE network for Public Safety services, or network operators deploying a dedicated LTE network for Public Safety services. This book is also intended for infrastructure and device vendors who plan to implement Public Safety features in their products and regulators who want to learn more about LTE and its use in the field of Public Safety. We hope everyone interested in the subject of this book benefits from the content provided.

December 2014, the Authors

Acknowledgments

The book has benefited from the extensive review of many subject matter experts and their proposals for improvements. The authors would like to thank, in particular, the following persons for their review and contribution: Gerald Görmer, Vesa Hellgren, Silke Holtmanns, Peter Leis, Zexian Li, Cassio Ribeiro, Mani Thyagarajan, Gabor Ungvari, Gyorgy Wolfner, Steven Xu, and Robert Zaus.

However, the authors are solely responsible for the content of this book.

They would also like to thank Sandra Grayson, Mark Hammond, Clarissa Lim, Liz Wingett, Lincy Priya & team from Wiley for their continuous support during the editing process.

Last but not least, the authors thank their families for their patience and cooperation rendered for writing the book.

The authors appreciate any comments and proposals for enhancements and corrections in future editions of the book. Feedback can be sent directly to rainer.liebhart@nsn.com, devaki.chandramouli@nsn.com, wong.curt@gmail.com, and juergen.merkel@nsn.com.

The authors would like to include additional thanks and full copyright acknowledgements as requested by the following copyright holders in this book.

© 2014, 3GPPTM. TSs and TRs are the property of ARIB, ATIS CCSA, ETSI, TTA and TTC who jointly own the copyright in them. They are subject to further modifications and are therefore provided here 'as is' for information purposes only. Further use is strictly prohibited.

Holma H. and Toskala A. (2009), 'LTE for UMTS OFDMA and SC-FDMA Based Radio Access'; JohnWiley & Sons, The Atrium, Southern Gate, Chichester, West Sussex, PO19 8SQ, UK.

Introduction

In a nutshell, this book explains how Long-Term Evolution (LTE) can be used as technology enabler for Public Safety networks, for example, how Public Safety networks can be built on top of LTE. We use the term LTE as synonym for the overall system that consists of radio and core networks, also known as Evolved Packet System (EPS). As a prerequisite, the book assumes some familiarity of the reader with basic concepts of mobile networks and especially with LTE. In Chapters 1 and 2, the book starts with an overview of LTE, its history, network architecture, and main features. Readers interested in detailed call flows for basic procedures should have a look into the Appendix. Chapters 1 and 2 are self-explanatory, but by nature they are not intending to give a full and complete overview of LTE. Hints for further reading, for example, consulting 3GPP specifications, are provided in the various chapters of the book. After reading Chapter 1, the reader should have a fairly good understanding of and why LTE was developed, knows the main architectural alternatives and used interfaces, knows the purpose of mobility management procedures, the LTE QoS concept, the purpose of bearers in LTE, and how they are established and has an understanding of LTE security. In addition, Chapter 1 describes features like Voice/SMS in LTE Multicast Broadcast Multimedia System (MBMS) and Network Sharing that are directly or indirectly of relevance for Public Safety.

Focus of Chapter 2 is on regulatory features and priority services available in LTE. These include support for emergency services, support for public warning messages, lawful intercept, and enhanced multimedia priority service. As these features are potentially important also in case of Public Safety networks, we provide an overview in this chapter. Chapters 1 and 2 should give the reader a good understanding of the feature-richness of the LTE standard.

While Chapter 3 explains the special nature of Public Safety networks and why LTE was chosen as technology for next-generation Public Safety networks, Chapters 4 and 5 are at the heart of this book. Chapter 4 describes so-called Proximity Services and their impacts on LTE radio and system design. This feature enables two LTE devices to discover each other and directly communicate, that is, communicate without network coverage. Thus, it introduces a fundamentally new form of communication using the standardized LTE technology as up to now two LTE devices can only communicate using the network.

Chapter 5 explains how a group communication service on top of LTE can be implemented. Efficient communication within a group of devices is, apart from device-to-device communication, the second important service in a Public Safety network. Group communication in this respect deals with all aspects of providing the same content to many LTE devices at the same time. This service is mainly used by "Push to Talk (PTT)" applications where one group member talks at a certain time and the content is distributed to all other group members.

Chapter 5 requires some knowledge about MBMS that is provided in Chapter 1. The application using such capabilities is up to now not specified in 3GPP or other organizations. However, 3GPP has started working on a so-called Mission Critical Push to Talk (MCPTT) application in its Release 13.

In Chapter 6, we give some hints about our expectations of this recently started activity on MCPTT. In addition, Chapter 6 explains why we think LTE was the right choice as technology for future Public Safety networks and gives an outlook to upcoming work in 3GPP.

Finally, the Appendix provides details for the most important call flows regarding mobility management, session management, MBMS procedures, and an overview of 3GPP reference points used throughout the book.

Terminology

2G	2nd Generation
3G	3rd Generation
3GPP	3rd Generation Partnership Program
AAA	Authentication, Authorization, and Accounting
ACB	Access Class Barring
ACBL	Access Class Barring List
ACBT	Access Class Barring Time
ACMA	Australian Communications and Media Authority
ADMF	Administration Function
AF	Application Function
AKA	Authentication and Key Agreement
ALUID	Application Layer User ID
AM	Acknowledged Mode
AMBR	Aggregate Maximum Bit Rate
AN	Access Network
ANDSF	Access Network Discovery and Selection Function
AP	Application Protocol, Access Provider
APCO	Association of Public Safety Communications Officials
APF	Access Probability Factor
API	Application Programming Interface
APN	Access Point Name
APN-AMBR	APN Aggregate Maximum Bit Rate
Apps	Applications
ARIB	Association of Radio Industries and Businesses
ARIB	Association of Radio Industries and Businesses
ARP	Address Resolution Protocol, Allocation and Retention Priority
AS	Access Stratum, Application Server
ATIS	Alliance for Telecommunications Industry Solutions
AuC	Authentication Center
AVP	Attribute Value Pair
BCH	Broadcast Channel
BCCH	Broadcast Control Channel
BD	Billing Domain
BM-SC	Broadcast Multicast Service Center

BSC	Base Station Controller
BSR	Buffer Status Reports
BSS	Base Station System
BTS	Base Transceiver Station
CAPEX	Capital Expenditure
CBC	Cell Broadcast Center
CBE	Cell Broadcast Entity
CBS	Cell Broadcast Service
CC	Content of Communication
CCSA	China Communication Standards Association
CDF	Charging Data Function
CDMA	Code Division Multiple Access
CDR	Charging Data Record
CGF	Charging Gateway Function
CI	Cell Identity
CMAS	Commercial Mobile Alert System
CN	Core Network
CP	Control Plane
CS	Circuit Switched
CSFB	Circuit Switched Fall Back
CTF	Charging Trigger Function
D2D	Device-to-Device
DF	Delivery Function
DHCP	Dynamic Host Configuration Protocol
DIAMETER	Pun on the name RADIUS (diameter is twice the circle radius)
DL	Downlink
DM	Device Management
DNS	Domain Name System
DPI	Deep Packet Inspection
DRX	Discontinuous Reception
DSCP	DiffServ Code Point
DSMIP	Dual Stack Mobile IP
DTLS	Datagram Transport Layer Security
E2E	End-to-End
ECRIT	Emergency Context Resolution with Internet Technologies
EDGE	Enhanced Data Rates for GSM Evolution
eHRPD	Evolved High Rate Packet Data
EIR	Equipment Identity Register
EEA	EPS Encryption Algorithm
EHPLMN	Equivalent Home PLMN
eMBMS	Evolved MBMS
eNB, eNodeB	E-UTRAN NodeB, also referred to as Evolved NodeB
ePDG	Evolved Packet Data Gateway
ECM	EPS Connection Management
EPS	Evolved Packet System
EMM	EPS Mobility Management

EPUID	EPC-level User ID
ETSI	European Telecommunications Standards Institute
ETSI TCCE	ETSI Technical Committee TETRA and Critical Communications Evolution
ETWS	Earthquake and Tsunami Warning System
E-UTRAN	Evolved Universal Terrestrial Radio Access Network
EU-ALERT	European Public Warning System
FBC	Flow-Based Charging
FCC	Federal Communications Commission
FDD	Frequency Division Duplex
GBA	Generic Bootstrapping Architecture
GBR	Guaranteed Bit Rate
GCS AS	Group Call System Application Server
GCSE	Group Call System Enablers
GERAN	GSM/EDGE Radio Access Network
GGSN	Gateway GPRS Support Node
GMK	Group Master Key
GO	Group Owner
GPRS	General Packet Radio Service
GPS	Global Positioning System
GSA	Global Mobile Suppliers Association
GSK	Group Session Key
GSM	Global System for Mobile Communications
GSMA	GSM Association
GSM-R	GSM for Railways
GTP	GPRS Tunneling Protocol
GUTI	Globally Unique Temporary Identity
GUMMEI	Globally Unique Mobile Management Entity Identifier
GW	Gateway
HARQ	Hybrid Adaptive Repeat and Request
HI	Handover Interface
HLR	Home Location Register
HPLMN	Home Public Land Mobile Network
HSDPA	High-Speed Downlink Packet Access
HSPA	High-Speed Packet Access (HSDPA+HSUPA)
HSUPA	High-Speed Uplink Packet Access
HSS	Home Subscriber Server
HTTP	Hyper Text Transfer Protocol
IC	Incident Commander
I-CSCF	Interrogating Call Session Control Function
IEC	Immediate Event Charging
IEEE	Institute of Electrical and Electronics Engineers
IETF	Internet Engineering Task Force
IMEI	International Mobile Equipment Identity
IMS	IP Multimedia Subsystem
IMSI	International Mobile Subscriber Identity

IOPS	Isolated E-UTRAN Operation for Public Safety
IP	Internet Protocol
IP-CAN	IP Connectivity Access Network
IPSec	IP Security
IRI	Intercept-Related Information
KMS	Key Management System
KPAS	Korean Public Alert System
L1	Layer 1
L2	Layer 2
L3	Layer 3
LA	Location Area
LAI	Location Area Identity
LBI	Linked EPS Bearer Identity
LEA	Law Enforcement Agency
LEMF	Law Enforcement Monitoring Facility
LIF	Location Interoperability Forum
LIPA	Local IP Access
LMA	Local Mobility Anchor
LMR	Land Mobile Radio
LTE	Long-Term Evolution
M2M	Machine-to-Machine
MAC	Medium Access Control
MAG	Mobile Access Gateway
MBMS	Multimedia Broadcast/Multicast Service
MBR	Maximum Bit Rate
MBSFN	MBMS Single Frequency Network
MCH	Multicast Channel
MCCH	Multicast Control Channel
MCE	Multi-cell Coordination Entity
MCPTT	Mission Critical Push To Talk
ME	Mobile Equipment
MGCF	Media Gateway Control Function
MIB	Master Information Block
MIC	Message Integrity Check
MIP	Mobile IP
MLP	Mobile Location Protocol
MME	Mobility Management Entity
MO	Mobile-Originated, Management Object
MSC	Mobile Switching Center
MSI	MCH Scheduling Information
MSISDN	Mobile Subscriber ISDN Number
MSP	MCH Scheduling Period
MT	Mobile Terminated
MTCH	Multicast Traffic Channel
NAS	Network Access Server (DIAMETER application)
NAS	Non Access Stratum

NB	NodeB
NPSBN	National Public Safety Broadband Network
NPSBN-U	NPSBN User
OA&M	Operation, Administration, and Maintenance
OCS	Online Charging System
OFCS	Offline Charging System
OFDM	Orthogonal Frequency Division Multiplexing
OMA	Open Mobile Alliance
OMA DM	OMA Device Management
OPEX	Operational Expenditure
OTA	Over-the-Air
OTT	Over-the-Top
P.25	Project 25
P2P	Peer-to-Peer
PAM	Priority Alarm Message
PCC	Policy and Charging Control
PCEF	Policy Charging Enforcement Function
PCI	Physical Cell ID
PCRF	Policy and Charging Rules Function
P-CSCF	Proxy Call Session Control Function
PD2DSS	Primary D2D Synchronization Signal
PDCP	Packet Data Convergence Protocol
PDN	Packet Data Network
PDP	Packet Data Protocol
PDU	Protocol Data Unit
PEK	ProSe Encryption Key
PFID	ProSe Function ID
P-GW	Packet Data Network Gateway (PDN-GW)
PGK	ProSe Group Key
PLMN	Public Land Mobile Network
PMCH	Physical Multicast Channel
PMIP	Proxy Mobile IP
PMK	Pairwise Master Key
PoA	Point of Attachment
PoC	Push To Talk over Cellular
ProSe	Proximity Services
PS	Packet Switched, Public Safety
PSAP	Public Safety Answering Point
PSS	Primary Synchronization Signal
PSSID	Physical Layer Identity
PSTN	Public Switched Telephone Network
PSCH	ProSe Communication Shared Channel
PTCH	ProSe Communication Traffic Channel
PTK	ProSe Traffic Key
PubS	Public Safety
PWS	Public Warning System

QCI	QoS Class Identifier
QoS	Quality of Service
RA	Routing Area
RAB	Radio Access Bearer
RADIUS	Remote Authentication Dial-In User Service
RAN	Radio Access Network
RAT	Radio Access Type
Rel	Release
RLC	Radio Link Control
RNC	Radio Network Controller
RRC	Radio Resource Control
RTCP	Real-Time Control Protocol
RTP	Real-Time Transport Protocol
SA	Scheduling Assignment
SAE	System Architecture Evolution
S-CSCF	Serving Call Session Control Function
SCTP	Stream Control Transmission Protocol
SD2DSS	Secondary D2D Synchronization Signal
SDF	Service Data Flow
SDP	Session Description Protocol, Service Delivery Platform
SGSN	Serving GPRS Support Node
S-GW	Serving Gateway
SIB	System Information Broadcast
SIM	Subscriber Identity Module
SIP	Session Initiation Protocol
SIPTO	Selective IP Offload
SMS	Short Message Service
SMSC	Short Message Service Center
SLA	Service Level Agreement
SLP	SUPL Location Platform
SR	Scheduling Request
SRTCP	Secure RTCP
SRTP	Secure RTP
SSID	Service Set Identifier
SSS	Secondary Synchronization Signal
TA	Tracking Area, Timing Advance
TAI	Tracking Area Identity
TAU	Tracking Area Update
TC MSG	Technical Committee Mobile Standards Group
TCO	Total Cost of Ownership
TCP	Transmission Control Protocol
TDD	Time Division Duplex
TE	Terminal Equipment
TEID	Tunnel Endpoint Identifier
TETRA	Terrestrial Trunked Radio
TFT	Traffic Flow Template

TIA	Telecommunications Industry Association
TM	Transparent Mode
TMGI	Temporary Mobile Group Identity
TMSI	Temporary Mobile Subscriber Identity
TR	Technical Report
TS	Technical Specification
TTA	Telecommunications Technology Association
TTC	Telecommunication Technology Committee
TWAG	Trusted WLAN Access Gateway
TWAN	Trusted WLAN Access Network
UDP	User Datagram Protocol
UE	User Equipment
UICC	Universal Integrated Circuit Card
UL	Uplink
UM	Unacknowledged Mode
UMTS	Universal Mobile Telecommunications System
UP	User Plane
URI	Uniform Resource Identifier
USA	United States of America
USIM	Universal Subscriber Identity Module
UTC	Universal Time, Coordinated
UTRAN	Universal Terrestrial Radio Access Network
v4	Version 4
v6	Version 6
VPLMN	Visited Public Land Mobile Network
WCDMA	Wideband Code Division Multiple Access
WLAN	Wireless Local Area Network
WLCP	WLAN Control Protocol
WLLID	WLAN Link Layer ID
WRC	World Radio Conference

1

Introduction to LTE/SAE

1.1 Role of 3GPP

The 3rd Generation Partnership Project (in short 3GPP) is a joint international standardization initiative between North American (Alliance for Telecommunications Industry Solutions (ATIS)), European (European Telecommunications Standards Institute (ETSI)), and Asian organizations (Association of Radio Industries and Businesses (ARIB) and Telecommunication Technology Committee (TTC) in Japan, Telecommunications Technology Association (TTA) in Korea and China Communication Standards Association (CCSA) in China) that was originally established in December 1998. The participating organizations are also called organizational partners. Scope of 3GPP was to specify a new worldwide mobile radio system (the Global System for Mobile Communications (GSM) was a European initiative while Code Division Multiple Access (CDMA) was initiated in North America, both are not compatible with each other) based on the evolved GSM techniques General Packet Radio Service (GPRS)/EDGE. This activity has led to the standardization of the third-generation Universal Mobile Telecommunications System (UMTS), which consists of Wideband Code Division Multiple Access (WCDMA) as radio technology and a core network supporting both circuit-based voice calls and packet-based data services. UMTS was meant as a universal standard that allows subscribers to use their UMTS-capable mobile phones and subscriptions worldwide through roaming (for an explanation of the term "roaming," see Section 1.13.1) agreements between mobile operators. UMTS is a big success story with around 1.4 billion WCDMA subscriptions deployed until now.

But 3GPP did not stop work after UMTS, in the following years enhancements of UMTS like High-Speed Packet Access (HSPA)/HSPA+, new services such as Multicast/Broadcast delivery, Location services, and the IP Multimedia Subsystem (IMS) were introduced. Long-Term Evolution (LTE) with a new Orthogonal Frequency-Division Multiplexing (OFDM)-based radio technology and an All-IP core network architecture is the newest development of 3GPP.

3GPP is organized in different working groups (see Figure 1.1) that are responsible for different parts of the 3GPP system. The Radio Access Network (RAN) groups define the radio parts of the UMTS/LTE system, i.e. the physical layer, and radio protocols. The GSM/EDGE Radio Access Networks (GERAN) groups work specifically on the maintenance and development of GSM/EDGE access technologies. The System Architecture (SA) and Core/Terminal

LTE for Public Safety, First Edition. Rainer Liebhart, Devaki Chandramouli, Curt Wong and Jürgen Merkel.
© 2015 John Wiley & Sons, Ltd. Published 2015 by John Wiley & Sons, Ltd.

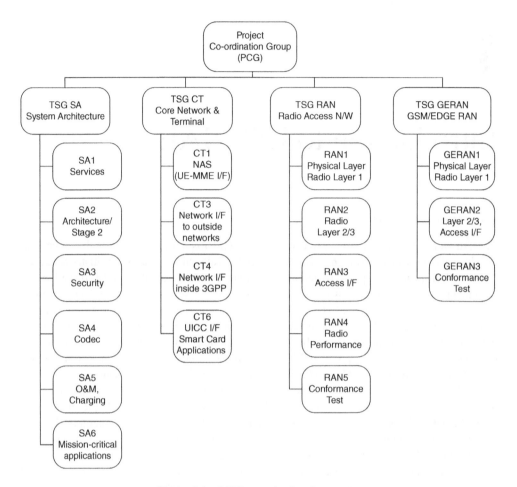

Figure 1.1   3GPP organizational structure

(CT) groups specify all parts of the overall system (e.g., architecture, security, charging) and all non-radio protocols (between the mobile device and network, within the network and between networks). A new working group SA6 will be operational from January 2015 onwards to standardize a Mission Critical Push To Talk (MCPTT) application in 3GPP. For details regarding MCPTT, please refer to Chapter 5.

3GPP follows a phased approach; working output is delivered as a set of Technical Specifications (TS) in so-called System Releases. Technical Specifications contain normative requirements that have to be implemented by chipset, device, and network equipment vendors. Interim results of ongoing work in 3GPP are usually captured in non-normative Technical Reports (TR). Test specifications are also created by 3GPP (mainly test cases for User Equipment (UE) to network communication). It has to be noted that 3GPP defines only functions and protocols, how these functions are implemented in concrete network nodes or whether some functions are implemented in the same node is up to the network vendor. One basic design principle in 3GPP's standardization process is backward compatibility of new features with existing ones. This ensures that new features can be introduced in one network without the need to upgrade all interconnected networks or all other nodes within this network at the same time.

Table 1.1   3GPP milestones up to Release 12

Release	Date	Main content
Phases 1 and 2	1992 and 1995	Basic GSM functions
Release 96, 97, 98, 99	1996, 1997, 1998, 1999	GPRS, HSCSD, EDGE, UMTS
Release 4	2001	MSC server split architecture
Release 5	2002	HSDPA, IMS
Release 6	2004	HSUPA, MBMS, Push to Talk over Cellular (PoC)
Release 7	2007	HSPA, EDGE evolution
Release 8	2008	LTE/SAE
Release 9	2009	LTE/SAE enhancements, Public Warning System (PWS), IMS emergency sessions over LTE/HSPA
Release 10	2011	LTE Advanced Local IP Access (LIPA) Selective IP Traffic Offload (SIPTO)
Release 11	2012	Heterogeneous Network (HetNet) Support Coordinated Multipoint Operation (CoMP)
Release 12	2014	Public Safety Machine type communication HSPA/LTE carrier aggregation

Table 1.1 provides a brief overview of the official release dates and milestones of the 3GPP releases up to Release 12.

For more information on the history and structure of 3GPP, visit the official 3GPP site at http://www.3gpp.org/about-3gpp/about-3gpp.

1.2   History of LTE

Main drivers for evolution of mobile networks are usually higher bandwidth on the air interface and better spectral efficiency (i.e., the information rate transmitted over a given bandwidth). After improving these key factors for WCDMA over several years, which led to the specification of HSPA and its evolution HSPA+, the 3GPP standardization forum, in 2004, started evaluating a new radio technology as successor for WCDMA. Objectives for starting this work were higher peak data rates (>100 Mbit/s in Downlink and >50 Mbit/s in Uplink) and lower latency besides other improvements. This work is formed under the name LTE. As the search for a more appealing name/acronym had no result, LTE is now used as the radio interface name in most official publications. Inside 3GPP the newly developed radio access network is called Evolved UMTS Radio Access Network (E-UTRAN) to indicate the evolution path from GERAN (GSM/GPRS/EDGE) second-generation networks (2G) to UTRAN (WCDMA/HSPA) third-generation networks (3G), and finally to E-UTRAN (LTE) fourth-generation networks (4G). It has to be noted that LTE initially did not fulfill International Telecommunication Union's (ITU) International Mobile Telecommunication (IMT) Advanced requirements, namely, the 1 Gbit/s peak data rate, which are officially the criteria for a network technology to be called 4G. Only LTE Advanced (LTE-A) will be able to support such data rates. So, strictly speaking calling LTE 4G was not correct but due to marketing battles in the United States, 4G was anchored as synonym for LTE and the ITU took the decision to allow any technology that provides an evolution path toward IMT Advanced to be called 4G. In parallel to the work on a new radio interface, 3GPP initiated a study to evolve the 2G/3G packet core network to which GERAN and UTRAN are connected (known as GPRS core) in order to cope with the new demands of LTE. This core network study was called System Architecture Evolution (SAE) and it was documented in the Technical Report TR 23.882 [1]. The final outcome of this work was a new packet core design in 3GPP's Release 8 documented in Technical Specifications TS 23.401 [2] and TS 23.402 [3], called the Evolved Packet Core (EPC); GPRS-specific parts are documented in TS 23.060 [4]. 3GPP Release 8 was officially completed in March 2009. EPC allows connecting LTE, GERAN/UTRAN, non-3GPP access systems such as Wireless Local Area Network (WLAN), WiMAX, and CDMA and also 3GPP-compliant small Femto Access Points (GERAN/UMTS/LTE radio stations connected via consumer links such as DSL or TV cable to the EPC) installed at homes, offices, and smaller campus areas. Special emphasis was put on optimized handover procedures between LTE and CDMA2000 eHRPD (Evolved High-Rate Packet Data) access due to requirements from CDMA network operators in the United States and Japan, which introduced LTE very early (starting in 2010). The EPC together with the connected radio access systems and the UEs is called Evolved Packet System (EPS) – the term LTE/SAE is also being used in this context. In contrast to the 2G and 3G systems, EPS no longer contains a Circuit-Switched (CS) part offering classical telephony network connectivity (e.g., EPS lacks optimized and dedicated radio bearers for CS voice calls) but only contains the Packet-Switched (PS) part providing data connectivity (for a description of CS and PS, please refer to Section 1.13.2). For that reason, supporting voice in EPS and providing a smooth migration story from 2G/3G voice and SMS (Short Message Service) to voice and SMS in EPS is extremely important for the acceptance of the new system. A lot of effort was spent in specifying solutions for the above-mentioned issues during Releases 8 and 9 (and continuing in later releases). This resulted in features called CS Fallback, SMS over SGs, SMS in MME (Mobility Management Entity), IMS Centralized Services, Session Continuity, and Single Radio Voice Call Continuity (SRVCC) among others.

As already mentioned EPC is an evolution of the 3GPP system architecture that finally realized the vision of an all-IP network. EPC in conjunction with the IMS delivers various services such as VoIP (Voice over Internet Protocol), Short Message Service (SMS), Video call, Picture share, Instant Messaging, and Presence. EPC and IMS support mobility with the existing 2G/3G wireless networks as well as fixed networks to facilitate smooth migration, interworking, and service continuity across all these networks. Nevertheless, as a pure IP-based network the main application for LTE/SAE will be the "Internet" with its rapidly increasing demands for more bandwidth and lower latency coming from P2P services or applications such as online gaming, Mobile TV, and Machine-to-Machine network deployments like smart traffic control or smart grid.

LTE/SAE was designed to cope with the challenges of the growing broadband mobile market: more data per user, "always-on" with high expectation on quality and reliability, connected everywhere, network efficiency, more devices that are not operated by humans, and the need to connect with an all-IP world.

As of today more than 300 operators in more than 100 countries have commercially launched LTE services. Nearly 160 million LTE subscriptions were issued worldwide, while nearly 1900

LTE capable devices were launched in the meantime. Without doubts LTE is the de facto standard in mobile broadband communication around the globe that will deliver broadband multimedia services to hundreds of million subscribers in the near future. These services, whether provided in the Internet or mobile networks, will be accessible with standardized devices (Smartphones, tablets, laptops) in nearly all countries of the world with an end user quality comparable to fixed broadband networks.

1.3 Drivers for LTE

As outlined in the previous section, main drivers to start work on a new radio technology and core architecture such as LTE/SAE are the need for higher data rates and a significant reduction of control plane latency and round trip delay to support future broadband, high quality services. After work on LTE started, 3G standards also progressed further, currently providing peak data rates well above 300 Mbit/s. Higher data rates are a must when recognizing the tremendous increase of mobile data traffic: forecasts indicated that mobile data traffic will increase 18-fold between 2011 and 2016 (an increase three times faster than fixed IP traffic) and will account for around 60% of the total IP traffic by 2016. As the usage of Internet services such as Email, browsing, chatting, or community applications increases dramatically in mobile networks, the limitations of 2G/3G radio and packet core networks has become apparent nowadays. Reduced control plane latency is a need to provide a high-quality (minimum delays when connecting to the network or during handover) "always-on" experience to the end user. Latency of control plane messages and big round trip delays were seen as drawbacks of existing legacy 2G/3G systems by many operators. Another driver for a flat, pure IP-based and simplified architecture with less radio and core network nodes was certainly the desire to decrease overall costs (OPEX and CAPEX). Finally, a purely IP-based architecture provides the possibility for introducing a PS optimized system while, for example, the 3G system had to make some compromises in support of packet-based services as 3G radio bearers have to be optimized for CS voice calls.

Another important design principle of EPC was backward compatibility and the capabilities to connect to non-3GPP access systems such as WLAN, WiMAX, or CDMA2000 systems. As such EPC and in general EPS are providing inherent mechanisms to support mobility for devices when changing radio access between 2G, 3G, and LTE, either based on GPRS Tunneling Protocol (GTP) or Proxy Mobile IP (PMIP) mobility protocols. For non-3GPP access systems such as WLAN or WiMAX, LTE/SAE is supporting mobility by reuse of generic mobility protocols defined in IETF, namely, PMIPv6 as specified in IETF RFC 5213 [5] and DSMIPv6 (Dual-Stack Mobile IP) as specified in IETF RFC 5555 [6]. For CDMA2000 eHRPD handover to LTE was improved by introducing special control plane and user plane interfaces to the EPC in order to deliver information from one access system to the other before the actual handover takes place to speed up the overall handover process. This harmonized core network architecture supporting 2G/3G, LTE, and non-3GPP access systems was another important objective when designing EPC.

While cost reduction is the main driver for network operators to introduce LTE/SAE, delay optimizations (minimized latency and round-trip delay leads to high-TCP traffic throughput, low-UDP/RTP traffic jitter to high-quality real-time services) and fast service availability caused by low bearer setup times are the benefits for the end user. These benefits increase the acceptance of the new technology and will pay back the investments of operators, terminal, and infrastructure vendors.

LTE/SAE is a broadband, standardized, wireless and packet-based system and constantly evolving to meet the needs of industry demands through 3GPP standardization activities. This allows public safety personnel to take advantage of an advanced wireless packet system and the vast support of innovative new applications for real-time information sharing and collaboration during emergencies and day-to-day operations. It improves situational awareness and enhances the safety of first responders and the public in general. LTE/SAE-based public safety networks will allow us to use any kind of multimedia services, voice, video, text, picture/file sharing, location-based services, in any situation (mobile and stationary) with high quality and in a reliable/resilient way. Built-in features such as network sharing and broadcast message delivery can be easily reused in public safety network deployments. Adopting IMS to provide mission critical push to talk and any kind of multimedia services over LTE is a further step toward a fully standardized, interoperable LTE public safety network that allows easy interconnection with other mobile and fixed networks, providing economy of scale on a high level.

1.4   EPS compared to GPRS and UMTS

When comparing E-UTRAN with UTRAN or GERAN the obvious change (besides use of the radio technology OFDM) is that E-UTRAN knows only one network element, the so-called Evolved NodeB (eNodeB or eNB) while in UTRAN NodeB and Radio Network Controller (RNC) and in GERAN Base Transmitter Station (BTS) and Base Station Controller (BSC) exist. Main reasons for this simplified E-UTRAN architecture were reducing complexity, latency, and costs while increasing data throughput.

On the core network side the obvious difference between the GPRS core (i.e., the 2G/3G core network with Serving GPRS Support Node (SGSN) and Gateway GPRS Support Node (GGSN) functions) and EPC is the strict and built-in separation of control and user plane in EPC. While this is possible in GPRS as well by using the so-called Direct Tunnel feature, that is, by which the user plane traffic is directly tunneled from RNC to GGSN bypassing the SGSN, the EPC was designed in this way from the very beginning. The main reason for this separation of user data and control (signaling) traffic was the ability to dimension the infrastructure for user traffic differently from the infrastructure parts handling the control of the user traffic, making it easy to adapt to growing user traffic needs. There are two functional elements in the user plane, the Serving Gateway (S-GW) and Packet Data Network Gateway (P-GW or PDN-GW), and one additional element in the control plane, the MME. As a consequence, in minimum two (eNodeB and combined S-GW/P-GW) and in maximum three nodes (eNodeB, S-GW, and P-GW) are in the EPS user plane path when LTE/SAE is deployed (1–2 less than in 2G/3G). This comes with a higher degree of simplicity, higher throughput, and less latency. Besides support of legacy 2G/3G access systems and LTE, the EPC also supports non-3GPP access systems that can be trusted or non-trusted from the EPC point of view. Support means that EPC provides means to authenticate, authorize, and charge subscribers using these non-3GPP access systems; the user plane is securely routed to and through the EPC toward a Packet Data Network (PDN) like the Internet. Last but not the least, mobility between 3GPP and non-3GPP access systems is enabled with the mobility anchor located in the EPC (P-GW). This kind of enhanced support of interworking with non-3GPP access systems is not supported by classical 2G GPRS.

In 2G/3G, the UE can be attached to the network without having any Packet Data Protocol (PDP) context established (i.e., no IP address is assigned to the UE). This concept was changed with the introduction of LTE to provide "always-on" connectivity to the UE anytime it is registered with the network. In LTE, a default bearer is established when the UE performs initial attach (explained later in this chapter) and when the last bearer is deactivated, the UE is detached. Thus, by default the UE is assumed to have at least one bearer context with which it can send and receive data when it is attached to an LTE network.

One technical detail that is not obvious is the concept of network-initiated bearer establishment. It was introduced to 2G/3G rather late and it was more an exception than the usual procedure. Until then, a bearer/context establishment request was always initiated by the UE. However, in LTE/SAE network-initiated bearer establishment is the main mechanism to establish dedicated bearers. Nevertheless, a UE-initiated bearer resource request procedure still exists in LTE/SAE but is considered as an exception.

To reduce the number of signaling messages between UE and core network the concept of a Tracking Area (TA) list allocated to the UE was introduced. Each TA of the list consists of one or more cells. Different TA lists allocated to different UEs in one area reduce the probability of simultaneous TA updates when a huge number of UEs are moving from one TA to another at the same time (e.g., when the users are riding the same train). Furthermore, a TA list allocated to one UE naturally decreases the need of the UE to perform TA updates but it comes with the drawback of bigger paging areas (the concepts of TA updates and paging are described later in this chapter).

1.5   Spectrum Considerations

To make LTE a worldwide success story, the technology must be flexible, adopt, and adapt to spectrum requirements in different countries around the globe. For that purpose, LTE was designed such that it can be deployed in a large variety of frequency bands, bands that might be allocated for a mobile broadband system by the national authorities in a specific country. Table 1.2 is taken from 3GPP TS 36.104 [7] and shows the supported Frequency Division Duplex (FDD) and Time Division Duplex (TDD) frequency bands for LTE.

More frequencies might be adopted for LTE once a commercial need arises in a specific country. It has to be noted that some bands are currently occupied by other technologies but LTE can coexist with these.

The process to allocate LTE spectrum for Public Safety deployments is ongoing. As it has not only technological but also economic and commercial consequences to allocate a certain portion of spectrum to a specific technology like LTE, spectrum-related discussions are ongoing in different regions of the world. However, some countries already took decisions which spectrum to allocate for Public Safety networks based on LTE. In the United States, the middle class tax relief and job creation act of 2012 reallocated the 700 MHz D-Block spectrum to public safety and included additionally $7 billion in federal funding for a nationwide LTE network for first responders.

The Australian Public-Safety Communications Officials (APCO) public safety bulletin No. 16 advised to develop a national interoperable public safety mobile broadband network based on LTE technology in the 4.9 GHz band. The Australian Communications and Media Authority

Table 1.2   LTE FDD/TDD frequency bands

Band	Uplink (UL) frequency band BS receive, UE transmit (MHz)		Downlink (DL) frequency band BS transmit, UE receive (MHz)		Duplex mode
1	1920	1980	2110	2170	FDD
2	1850	1910	1930	1990	FDD
3	1710	1785	1805	1880	FDD
4	1710	1755	2110	2155	FDD
5	824	849	869	894	FDD
6	830	840	875	885	FDD
7	2500	2570	2620	2690	FDD
8	880	915	925	960	FDD
9	1749.9	1784.9	1844.9	1879.9	FDD
10	1710	1770	2110	2170	FDD
11	1427.9	1447.9	1475.9	1495.9	FDD
12	699	716	729	746	FDD
13	777	787	746	756	FDD
14	788	798	758	768	FDD
15	Reserved		Reserved		FDD
16	Reserved		Reserved		FDD
17	704	716	734	746	FDD
18	815	830	860	875	FDD
19	830	845	875	890	FDD
20	832	862	791	821	
21	1447.9	1462.9	1495.9	1510.9	FDD
22	3410	3490	3510	3590	FDD
23	2000	2020	2180	2200	FDD
24	1626.5	1660.5	1525	1559	FDD
25	1850	1915	1930	1995	FDD
26	814	849	859	894	FDD
27	807	824	852	869	FDD
28	703	748	758	803	FDD
29	N/A	N/A	717	728	FDD[2]
30	2305	2315	2350	2360	FDD
31	452.5	457.5	462.5	467.5	FDD
...					
33	1900	1920	1900	1920	TDD
34	2010	2025	2010	2025	TDD
35	1850	1910	1850	1910	TDD
36	1930	1990	1930	1990	TDD
37	1910	1930	1910	1930	TDD
38	2570	2620	2570	2620	TDD
39	1880	1920	1880	1920	TDD
40	2300	2400	2300	2400	TDD
41	2496	2690	2496	2690	TDD
42	3400	3600	3400	3600	TDD
43	3600	3800	3600	3800	TDD
44	703	803	703	803	TDD

Note 1: Band 6 is not applicable.
Note 2: Restricted to E-UTRA operation when carrier aggregation is configured.

(ACMA) allocated an additional 60 MHz of spectrum across a number of bands (e.g., 800 MHz spectrum) to facilitate the deployment of high-speed, nationally interoperable mobile broadband networks for use by Australia's public safety authorities.

The European community is planning to recommend spectrum for public safety use in the 400 or 700 MHz bands. European spectrum community officials stated that the World Radio Conference (WRC) to be held in November 2015 is the best opportunity to get a dedicated spectrum allocation for Public Safety LTE deployments.

In Europe, many Asia-Pacific countries and parts of South America, the 400 MHz band is currently used by public safety agencies for their Terrestrial Trunked Radio (TETRA) and TETRAPOL systems. Using the same frequency band for LTE-based public safety networks around the world will allow minimizing investment costs by reuse of existing sites and assets.

In summary, Map 1.1 shows example bands that may be used for Public Safety. These bands are listed in ITU-R Res 646 and are recommended for Public Safety use. Which bands will be actually used in which region depends on national regulation at the end.

1.6   Network Architecture

1.6.1   Radio Access Network and Core Network

A mobile network (also called Public Land Mobile Network – PLMN) is usually separated into a Radio Access Network (RAN) and a Core Network (CN). The functional elements are specified by 3GPP. Functional elements relevant for LTE are described in later sections and chapters of this book (see e.g., section 1.6.5).

The RAN consists of all functions that are necessary to establish, maintain, and teardown connections between a user device (also called UE) and the network via the air interface. The RAN consists of radio base stations with their antennas that are spread over the geographical serving areas of the whole country.

The CN consists of all functions that are necessary to authenticate and authorize the user, setup voice calls or data connections, support mobility, charging, and lawful interception. The CN provides also interfaces to other mobile or fixed networks and to data networks such as the Internet or a company Intranet. Subscriber data are stored in a central register called Home Subscriber Server (HSS) that is part of the CN.

1.6.2   Architecture Principles

LTE/SAE has evolved from the 2G/3G PS domain, and especially EPC has its roots in the GPRS core network. Separation of control and user plane functions was a key in the design of EPC, thus the EPC basically consists of three functional elements. One is the MME that resides in the control plane of EPC. The MME can be seen as an evolution of the SGSN control plane function in GPRS. The Serving Gateway (S-GW) correlates with the SGSN user plane function in GPRS. All user plane packets in Uplink (UL) and Downlink (DL) are traversing the S-GW and the S-GW also acts as a local mobility anchor that is able to buffer downlink packets during handover. The P-GW finally is the global IP mobility anchor point comparable to

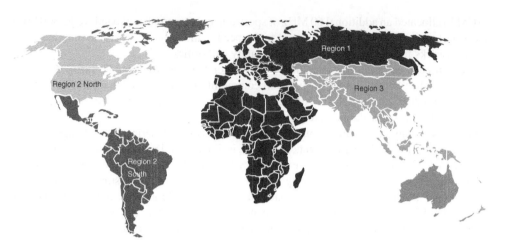

Map 1.1 Potential frequency bands for Public Safety. Numbers indicate MHz. Region 1 (mainly Africa, Europe, and Russia): 380–385/390–395. Region 2 North (United States and Canada): 3GPP band 14, 758–768/788–798. Region 2 South (Latin America): 746–806, 806.869, 4940–4990. Region 3 (mainly Asia): 406.1–430, 440–470, 806–824/851–869, 4940–4990, 5850–5925. Australia and other countries plan for the APT700 band (APT = Asia-Pacific Telecommunity), which is a segmentation of the 698–806 MHz band

the GGSN in GPRS. It allocates IP addresses to UEs and provides the interface toward Packet Data Networks (PDNs) such as the Internet or the mobile operator's service domain. The P-GW also contains the Policy Enforcement Function (PCEF) for the detection of service data flows, policy enforcement (e.g., discarding of packets), and charging (see section 1.6.12). All these network elements are logical functions, that is, in real implementations two or more functions (e.g., S-GW and P-GW) can reside on the same physical hardware platform. While the MME is selected by the eNodeB for a new session, the MME itself selects S-GW and P-GW by constructing special domain names and resolving these names by means of operator's Domain Name System (DNS) infrastructure. In the following sections, we will describe the EPS architecture variants for non-roaming and roaming cases and the architectures for interworking to 2G/3G. The functional description of the network elements can be found in Section 1.6.5.

1.6.3   Non-roaming Architecture

Figure 1.2 gives an overview of the logical LTE/SAE architecture in the non-roaming case, that is, the UE is served by its Home Public Land Mobile Network (HPLMN). In this and the following architecture figures, control plane interfaces are indicated with dotted lines while user plane interfaces are using full solid lines.

An overview of the main EPS reference points, their roles, and the underlying protocols can be found in the Appendix.

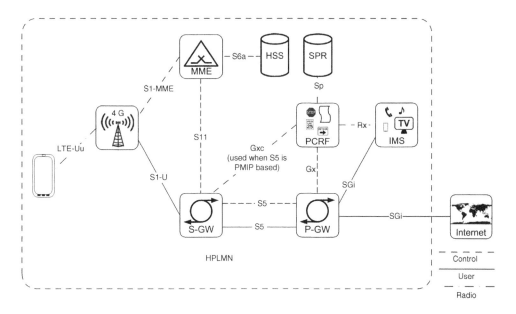

Figure 1.2   LTE/SAE non-roaming architecture

1.6.4   Roaming Architectures

1.6.4.1   Home Routed Roaming Architecture

The roaming architecture, that is, the UE is served by a VPLMN, is not much different from the non-roaming architecture. Main difference between non-roaming and roaming architecture is that in the roaming case S-GW is located in the VPLMN while P-GW is usually located in the HPLMN as most of the traffic is home routed traffic. Only if local breakout is used, the P-GW can be located in the VPLMN, but this scenario requires special arrangements between operators (e.g., usage of special Access Point Names (APN), providing charging information from VPLMN to HPLMN) and the user must be subscribed to this service. Thus, home routed traffic is the dominant usage scenario for PS services until now and this may be true also in the near future. The roaming architecture in the home routed scenario is shown in Figure 1.3.

As can be seen the S5 interface is replaced by S8 in roaming cases and in fact both are providing nearly identical functionality. This is similar to usage of Gn/Gp interfaces (see TS 29.060 [8]) in 2G/3G. S8 is either based on GTP or PMIP like S5. However, using PMIP for S8 requires either appropriate roaming agreements between operators or the VPLMN needs an interworking function to translate PMIP into GTP at the network border.

1.6.4.2   Local Breakout Roaming Architecture

Finally, we show the roaming architecture for local breakout to use services accessible via the VPLMN. The key point in this scenario is that the P-GW is located in the VPLMN and

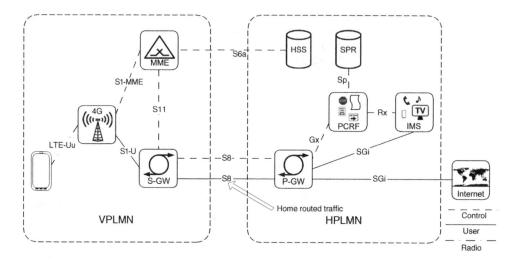

Figure 1.3   Roaming with home routed traffic

a Visited PCRF (V-PCRF) exists in the VPLMN as well to terminate the Gx interface (Gx is not an inter-operator interface). To receive QoS rules from HPLMN where the subscription data are stored (e.g., the subscribed maximum bit rate), a new interface S9 was introduced that connects the Home PCRF (H-PCRF) in HPLMN with the V-PCRF in VPLMN.

One possible solution to obtain VPLMN services requires the use of specially constructed APNs that are configured in the UE and used in the VPLMN to resolve them to a P-GW address in the VPLMN. Another possible solution is for UEs to use well-known standardized APNs like the IMS APN (see also Section 1.7) defined by Global System for Mobile Communications Association (GSMA) to get connectivity to a P-GW in the VPLMN.

As can be seen from Figure 1.4, the user can potentially use operator services in the HPLMN and VPLMN when local breakout is deployed.

1.6.5   Description of Functional Entities

1.6.5.1   User Equipment, Mobile Equipment, and the Universal SIM

In 3GPP terminology the UE is a device used by the end user for communication with the network. It is typically a smartphone, tablet, or modem equipped with LTE radio and is most often multiradio capable, that is, equipped also with other radios such as WLAN, cdma2000® 1xRTT, UTRAN, and GERAN. A UE consists of a Mobile Equipment (ME) part, which is the phone hardware that might be constituted of display, keypad, battery, and all electronics necessary to access and communicate with the network and providing the interface to the user. The only other piece of hardware required to form a UE is a Universal Integrated Circuit Card (UICC), which according to the definition in 3GPP TS 21.905 [9] is a "physically secure device, an IC card (or "smart card"), that can be inserted and removed from the terminal. It may contain one or more applications. One of the applications may be a USIM."

The Universal Subscriber Identity Module (USIM) contains all data and algorithms necessary to authenticate the UE and verifying the authenticity of the network (only in case of

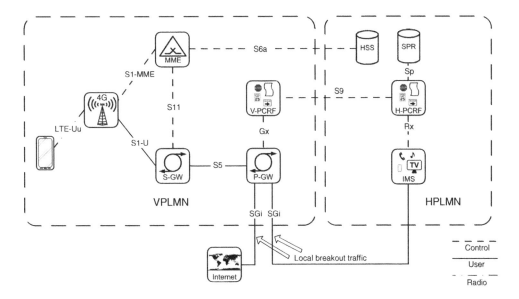

Figure 1.4   Roaming with local breakout

UMTS and LTE). Frequently, the terms UICC and USIM are referred to as Subscriber Iden-tity Module (SIM), as this is the term used in the predecessor organization of 3GPP, the ETSI Special Mobile Group (ETSI SMG).

In addition to the USIM application, the UICC can also contain other applications like the IMS – SIM (ISIM) that provides all necessary data for the UE to access the IP Multimedia Services domain.

A typical UE must go through a series of steps to request and manage a service. When it powers on it needs to perform the following steps:

1. Scanning for a LTE cell, synchronize with the network, and listen for system information over the broadcast channels.
2. Establish a signaling connection in order to communicate with the network.
3. Register with the network.
4. Establish a data connection to be "always-on."
5. Respond to authentication requests when initiated by the network.
6. Receive an IP address for IP connectivity.

Afterwards the UE can request for specific resources that are needed to run one or more applications.

1.6.5.2   E-UTRAN Node B (eNodeB/eNB)

Main part of the E-UTRAN architecture is the eNodeB. The name is derived from the NodeB in UMTS and just extended with the letter "e" that stands for "evolved." It is the base station that is in control of all radio-related functions. It is typically deployed throughout the network coverage area, each eNodeB resides near the actual radio antennas. As the eNodeB is the only E-UTRAN node it consists of functions that reside in UMTS in the NodeB and partly in the RNC (other RNC functions were moved to the MME).

It terminates the radio protocols from the UE and relays data between UE and EPC. It supports the following main functions:

- Ciphering/deciphering of user plane traffic, IP header compression, and decompression.
- Radio Resource Management (RRM) functions including radio bearer control and admission control.
- Scheduling traffic according to the assigned Quality of Service (QoS), constant monitoring of the resource usage situation, and radio resource allocation in both uplink and downlink directions toward the UE.
- Mobility management functions: controls and analyzes the radio signal level measurements carried out by the UE, performs measurements, and takes handover decisions for UE(s) based on current radio link quality.
- Select a MME and route the data traffic to the S-GW.

The eNodeBs are interconnected by the X2 interface, connected to the MME by the S1-MME (control plane) interface, and connected to the S-GW by the S1-U (user plane) interface. Note that an eNodeB can connect to multiple MME(s) and/or multiple S-GWs for load balancing and network sharing purposes.

1.6.5.3   Mobility Management Entity (MME)

The MME is the main control element in the EPC. It is derived from the control plane part of the 2G/3G SGSN, enhanced with some functions that were inherited from the 3G RNC. Typically a MME will be a server in a secure location in operator's premises. The MME is only part of the control plane path. User plane data bypass the MME. Thus, the MME has no charging functions, except when it supports the feature "SMS in MME" (see Section 1.8.2).

Besides physical connections to the eNodeB, the MME has also a logical connection with the UE. This logical connection is referred to as Non Access Stratum (NAS). A description of NAS and Access Stratum (AS) can be found in Section 1.13.3. The MME connects also to the user's home HSS to authenticate and authorize the UE and retrieve subscription data. In a nutshell, MME supports the following functions:

- Control plane traffic handling (termination of signaling from the UE).
- Session and mobility management, for example, idle mode mobility and handover control.
- Paging of UE(s) that are in idle mode.
- TA list management.
- Selection of P-GW and S-GW.
- MME selection during handover.
- Coordinates inter-S-GW and inter-MME relocations.
- Authentication and authorization of the UE.
- Bearer management including dedicated bearer establishment.
- Lawful interception of signaling traffic.

1.6.5.4   Serving Gateway (S-GW)

Main function of the Serving Gateway is routing user plane packets between E-UTRAN and P-GW. The S-GW is derived from the user plane part of the 2G/3G SGSN. It is part of the network infrastructure and can be deployed centrally or decentrally in the network. It acts as the local anchor point for inter-eNodeB mobility. If a bearer was established for a certain UE and the UE is in idle mode, that is, there is no signaling connection between UE and network, the S-GW will buffer incoming data packets and request the MME to page the UE. The S-GW is connected to one or more P-GW(s). For each UE and each bearer, a tunnel between S-GW and P-GW is established. The S-GW supports the following functions:

- User plane anchor for mobility between 2G/3G and LTE and for inter-eNodeB handover.
- Lawful Interception (LI).
- Packet buffering and initiation of paging.
- Packet routing and forwarding.
- Transport level packet marking.
- Generation of charging events for interoperator accounting in case of roaming.

1.6.5.5   Packet Data Network Gateway (P-GW)

The P-GW is the first IP hop router from UE point of view and the edge router between the EPS and external PDNs. It is the central mobility anchor and acts as the IP point of attachment for the UE. The UE may be connected to multiple PDNs at the same time through the same or different P-GWs (connecting to different P-GWs is in practice rather unusual). During handover the P-GW is not changed. It is anchoring the user plane for inter-S-GW mobility by maintaining tunnels per UE and bearer toward the S-GW. It is also part of the network infrastructure maintained centrally in operator premises.

One important function of the P-GW is the allocation of an IP address to the UE per PDN connection. On the basis of the particular PDN this can be a private or public IPv4 address or an IPv6 prefix.

On the basis of dynamic or static policy rules the P-GW performs enforcement functions such as shaping, gating, filtering, and packet marking. More precisely, enforcement is done in the Policy Enforcement Function (PCEF), which is an integral part of the P-GW. It ensures that the downlink data rate does not exceed the allowed (i.e., subscribed) maximum bit rate for a particular user and that data packets received in downlink direction are using the correct bearer/tunnel (this is called bearer binding). Following are the main functions supported by the P-GW:

- Edge router to other networks.
- Allocation of IP addresses to the UE.
- Central mobility anchor.
- Policy and Charging Enforcement, bearer binding.
- Packet Filtering (optionally Deep Packet Inspection).
- Accounting per UE and per bearer.
- Lawful Interception.

1.6.5.6 Policy and Charging Rules Function (PCRF)

The Policy and Charging Rules Function (PCRF) is responsible for dynamic policy and charging control (PCC). For more details on PCC, see Section 1.6.7. It is usually located in the operator premises along with other core network elements, for example, close to the P-GW.

The PCRF translates session data coming from the application layer (e.g., information on the codec used for a multimedia session) into access specific parameters. On the basis of these parameters it generates so-called PCC rules that specify which kind of QoS is applicable for which IP flows. These rules are installed and later on executed in the Policy Control Enforcement Function (PCEF) that is part of the P-GW. These rules also specify whether the P-GW should grant resource requests and whether it is allowed to process packets for a given IP flow. The PCRF takes subscription information (e.g., the maximum allowed bit rate) stored in the Subscriber Profile Repository (SPR) into account to make its decisions. The SPR is usually part of the HSS. In the 3GPP architecture the SPR has Sp interface to the PCRF, but this interface was never specified. If GTP is used between S-GW and P-GW, QoS rules are provided from PCRF to P-GW as P-GW can send QoS parameters further down to the RAN via GTP. If an operator decides to use PMIP between S-GW and P-GW, bearers are terminated in the S-GW as PMIP has no bearer concept implemented. In this case, QoS rules are provided from the PCRF to the S-GW while policy and charging rules are still provided to the P-GW.

1.6.5.7 Home Subscriber Server (HSS) and Subscriber Profile Repository (SPR)

The HSS is the central database for all subscriber related data in the network. It contains subscriber-specific data such as International Mobile Subscriber Identity (IMSI), Mobile Station International ISDN Number (MSISDN), subscribed APN, priority indication, and subscribed supplementary voice services (e.g., call forwarding). It is centrally located in the operator premises. The HSS contains logically the Home Location Register (HLR) function that is the central register in legacy 2G/3G networks. Originally the HSS function was introduced with 3GPP Release 5 to host IMS subscription data like the IMPI and IMPU(s).

Although the SPR is an extra functional entity defined by 3GPP, it is usually (in real-life implementations) collocated with the HSS. The SPR contains subscription information that is used by the PCRF to make policy decisions and generate appropriate PCC rules. Such data are, for example, the maximum allowed bit rates for a user, the subscribed guaranteed bandwidth, whether user has a prepaid or postpaid contract and whether user receives preferred treatment based on his status ("bronze," "silver," and "gold" users).

The HSS is involved in user authentication and authorization and other security-related functions, generating and storing keys, parameters for ciphering, mutual authentication, and message integrity checking. The HSS also records user's physical location (the addresses of MME, SGSN, and MSC serving the user). More than one HSS may be present in the network, depending on the number of subscribers and capacity of the hardware platform. In this case, identifiers such as MSISDN or IMSI can be used to select the correct HSS.

When the MME authenticates the user and authorizes the user request to access network resources (e.g., authorizing the APN provided by the UE), it needs access to the subscription data stored in the HSS. The HSS provides also the security key K_{ASME} from which the MME derives the security keys to cipher and integrity protect NAS messages exchanged between UE and MME. For location management purposes, the HSS stores the addresses of MME serving a particular UE.

1.6.5.8 Authentication, Authorization, and Accounting (AAA) Function

The Authentication, Authorization, and Accounting (AAA) server (not shown in the various architecture figures) is either used to authenticate and authorize users who are accessing the EPC through non-3GPP access systems or, optionally, by the P-GW to authorize usage of the provided APN and to allocate an IP address to the UE. The P-GW uses RADIUS or DIAMETER according to 3GPP TS 29.061 [10] to access the AAA server via the SGi interface. The operator's AAA server can also work as an AAA proxy and interwork with AAA servers of other PLMN(s) or in company networks. It is centrally located in the operator premises and has a direct interface to the HSS.

1.6.6 Session Management

1.6.6.1 Quality of Service and EPS Bearers

The EPS provides IP connectivity between a UE and a packet data network external to the PLMN. This is referred to as PDN connectivity service. An EPS bearer uniquely identifies traffic flows that receive a common QoS treatment. It is the level of granularity for bearer level QoS control in the EPC/E-UTRAN. All traffic mapped to the same EPS bearer receives the same bearer level packet forwarding treatment. Providing different bearer level packet forwarding treatment requires separate EPS bearers.

An EPS bearer is referred to as a GBR bearer, if dedicated network resources related to a Guaranteed Bit Rate (GBR) are permanently allocated once the bearer is established or modified. Otherwise, an EPS bearer is referred to as a non-GBR bearer.

Each EPS bearer is associated with a QoS profile including the following data:

- QoS Class Identifier (QCI): A scalar pointing in the P-GW and eNodeB to node-specific parameters that control the bearer level packet forwarding treatment in this node.
- Allocation and Retention Priority (ARP): Contains information about the priority level, the pre-emption capability, and the pre-emption vulnerability. The primary purpose of the ARP is to decide whether a bearer establishment or modification request can be accepted or needs to be rejected due to resource limitations.
- GBR: The bit rate that can be expected to be provided by a GBR bearer.
- Maximum Bit Rate (MBR): Limits the bit rate that can be expected to be provided by a GBR bearer.

Following QoS parameters are applied to an aggregated set of EPS bearers and are part of user's subscription data:

- APN Aggregate Maximum Bit Rate (APN-AMBR): Limits the aggregate bit rate that can be expected to be provided across all non-GBR bearers and across all PDN connections associated with the APN.

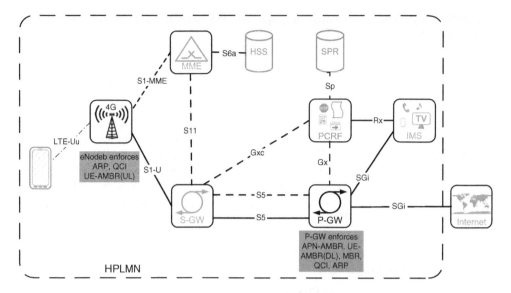

Figure 1.5   QoS enforcement in EPS

- UE Aggregate Maximum Bit Rate (UE-AMBR): Limits the aggregate bit rate that can be expected to be provided across all non-GBR bearers of a UE.

The UE routes uplink packets to the different EPS bearers based on uplink packet filters assigned to the bearers while the P-GW routes downlink packets to the different EPS bearers based on downlink packet filters assigned to the bearers in the PDN connection.

Figure 1.5 shows the nodes where QoS parameters are enforced in the EPS system.

1.6.6.2   Session and Bearer Management

With Session and Bearer Management procedures, EPS bearers for a particular UE are established and maintained. The default EPS bearer context is activated during the EPS Attach procedure. Upon successful attach, the UE can request setting up connections to additional PDNs. For each additional PDN connection, the MME activates a separate default EPS bearer. A default EPS bearer context remains activated throughout the lifetime of the PDN connection. Each PDN connection is characterized by the IP address that is assigned to this connection.

A dedicated EPS bearer is always linked to a default EPS bearer and inherits from it characteristics like the IP address, that is, UL traffic on a dedicated bearer uses the same source IP address as traffic on the default bearer. A dedicated bearer is used when additional EPS bearer resources with a specific QoS between UE and PDN are required. The distinction between default and dedicated bearers is transparent to the eNodeB. As an example: In case of VoLTE (see Section 1.8.1) voice and video streams use dedicated bearers associated with the IMS PDN connection while IMS/SIP signaling can use the default EPS bearer. A dedicated bearer is established via the dedicated bearer context activation procedure. It can either be part of the attach procedure or initiated together with the default EPS bearer context activation procedure.

The procedure is usually initiated by the network, but may be also requested by the UE. If the UE requests additional EPS bearer resources, the network decides whether to fulfill such a request by activating a new dedicated bearer or modifying an existing dedicated or default bearer.

By using the PCC framework the network can initiate the activation of dedicated EPS bearers together with the activation of the default EPS bearer or at any time later, as long as the default EPS bearer remains activated.

Default and dedicated EPS bearers can be modified. Dedicated EPS bearers can be released without affecting the default EPS bearer. If the default EPS bearer is released, all dedicated bearers linked to it are also released.

Readers interested in detailed call flows for UE-requested PDN connectivity request and network-initiated dedicated bearer activation procedure may refer to the Appendix.

1.6.6.3   IP Address Allocation

The network can assign three types of IP addresses to the UE once it connects to the network: either a private or public IPv4 address or an IPv6 prefix or both. The IP address can be allocated by the HPLMN, VPLMN (in case of local breakout), or potentially by an external service provider (e.g., a company network).

The type of IP address allocated to the UE depends on the PDN connection and on UE capabilities. The HSS stores one or more PDN types per APN in the subscription data. During Attach or UE-requested PDN connectivity procedure, the MME takes the requested and the subscribed PDN types into account before finally determining the PDN type for the connection.

DHCPv4 or 3GPP-specific signaling during PDN connection establishment is used to deliver an IPv4 address to the UE. IPv6 stateless address auto-configuration is used to assign a 64-bit globally unique prefix to the UE. If a shorter prefix has to be allocated for a PDN connection, it is delivered using DHCPv6 prefix delegation after the UE has first received the 64-bit prefix with stateless auto-configuration. It has to be noted that neither DHCPv4 nor DHCPv6 prefix delegation is currently widely used in live networks. In most of the cases, an AAA server assigns an IP address for a PDN connection and provides it to the P-GW via SGi signaling (RADIUS or DIAMETER).

1.6.7   Policy and Charging Control

The PCC system allows operators to dynamically maintain IP bearers with a certain QoS and provide means for online and offline charging of single-service data flows (for the charging aspects please refer to Section 1.6.12).

PCC helps to enforce the service data flows that are transmitted over a certain bearer. A service data flow is an aggregate of IP packet flows defined by 5-tuple filters (note that one service data flow can consist of several IP packet flows). A bearer can be seen as a transmission channel from the UE via eNodeB toward GGSN/P-GW with a certain capacity, delay, and bit error rate provided to all service data flows transported within it. Enforcement of QoS is performed hop-by-hop, on the radio link between UE and eNodeB, on the transport link between eNodeB and GGSN/P-GW, and finally beyond the GGSN/P-GW. However, support for QoS beyond GGSN/P-GW is out of scope of 3GPP and is up to configuration.

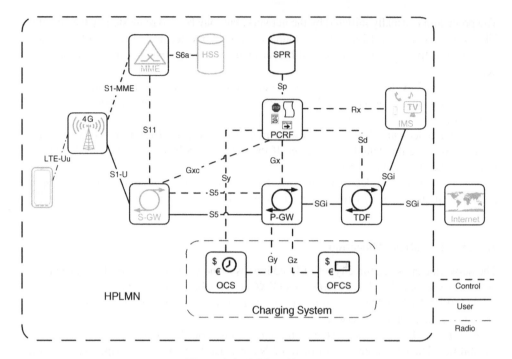

Figure 1.6 PCC architecture

Figure 1.6 gives a simplified overview of the PCC architecture. For details, see 3GPP TS 23.203 [11].

The central control function in the PCC architecture is the Policy and Charging Rules Function (PCRF). Basic idea is that the PCRF obtains information about new and ongoing media sessions from the application layer (called Application Function (AF)) and translates this into so-called policy rules that are enforced at the PCEF. AF and PCRF communicate via the DIAMETER-based Rx interface with each other (see 3GPP TS 29.214 [12]). In LTE networks the PCEF is implemented in the P-GW. The PCEF is responsible for the binding of service data flows to bearers, that is, to decide which flow goes into which bearer. IP flows in downlink and uplink that are traveling through the P-GW are then handled according to these policy rules. Policy rules are installed by the PCRF in the PCEF via the Gx interface, which is also based on DIAMETER. The PCC rule determines the authorized QoS, which is also signaled to the eNodeB, applicability of online/prepaid or offline/postpaid charging for this particular bearer, and addresses of respective online or offline charging servers. Policy rules can enforce traffic from/to certain destinations on a bearer, thus blocking traffic flows from/to other destinations. QoS-relevant subscription data (e.g., subscribed maximum bandwidth per user) are stored in the SPR, which is usually part of the HSS, but can be a stand-alone function. The PCRF has access to the SPR via the non-standardized Sp interface. The P-GW interacts with the Online and Offline Charging Systems (OCS/OFCS) via the DIAMETER-based Gy/Gz interfaces. Alternatively, Gz may use the GTP' protocol as well, which is a flavor of GTP, and P-GW may even send offline charging records directly to a billing system without passing through an OFCS. The OCS allows deploying prepaid charging schemes based on time or volume usage. The Sy interface allows the OCS to change policy rules based on the current usage of resources, for example, to downgrade the available bit rate for a user who has spent more than 4 GB data volume in a month. The Traffic Detection Function (TDF) is the 3GPP terminus for a Deep Packet Inspection (DPI) node and allows detecting applications based on different criteria like IP 5-tuples or the characteristics of application-specific messages. The PCRF can instruct the TDF via the Sd interface to start detection of applications for certain users and the TDF can inform the PCRF that an application was detected. On the basis of this information, policy rules can be installed at the PCEF (e.g., to block the service or to apply different charging rules). The TDF can be a stand-alone function or co-located with the P-GW.

1.6.8   Interfaces and Protocols in EPS

1.6.8.1   Control Plane

The control plane consists of all protocols used for signaling between UE and network or between two network entities.

UE – eNodeB – MME
Figure 1.7 shows the control plane protocol stack between UE, eNodeB, and MME.

A description for the 3GPP defined protocols (white background in the figure) and corresponding reference specifications are provided here. Other protocols based on IETF standards are not described.

Non Access Stratum (NAS)
NAS is the highest layer of the control plane between UE and MME at the radio interface. Main functions of the NAS protocol are the support of mobility of the UE and the support of session management procedures to establish and maintain IP connectivity between UE and P-GW. For further details, refer to 3GPP TS 24.008 [13], 3GPP TS 24.301 [14], and 3GPP TS 23.122 [15]. See also Section 1.13.3 for a comparison of NAS and AS.

Radio Resource Control (RRC)
RRC is the primary control protocol of the LTE–Uu interface, which is the interface between UE and eNB. It is responsible for a wide variety of system functions, including system information over the broadcast channel, radio configuration (set up, maintenance, and teardown of radio resources), measurements, and mobility. For further details, please refer to 3GPP TS 36.300 [16].

Packet Data Convergence Protocol (PDCP)
PDCP is responsible for ensuring the integrity of packets sent over the air interface. In addition, the PDCP layer compresses the IP header of a data packet. For further details, please refer to 3GPP TS 36.300 [16].

Radio Link Control (RLC)
RLC provides a logical link control mechanism over the air interface. Main task of RLC is the segmentation of PDCP packets in smaller parts that can be transmitted over the air. For further details, please refer to 3GPP TS 36.300 [16].

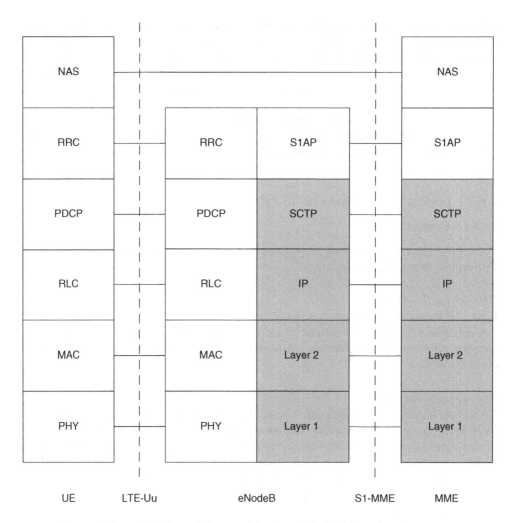

Figure 1.7   Control plane stack between UE, eNodeB, and MME

Medium Access Control (MAC)
The MAC layer selects a transport channel to transmit data and manages the mapping of logical channels to transport channels. In addition, the MAC layer is responsible to multiplex data to common and shared channels. For further details, please refer to 3GPP TS 36.300 [16].

Physical Layer (PHY)
This is the layer 1 of the LTE–Uu interface. Physical channels differentiate the source of the transmission as well as its destination. A particular message may be destined for a specific UE (a handover command) or may be intended for all active UE(s) (system broadcast messages). For further details, please refer to 3GPP TS 36.300 [16].

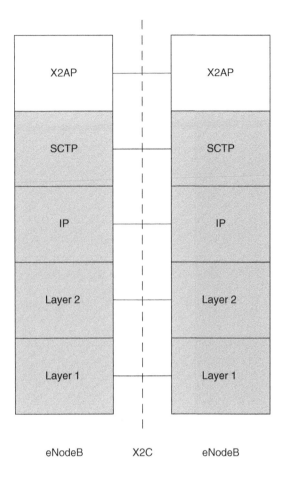

Figure 1.8   Control plane stack between eNB and eNB

S1-Application Protocol (S1AP)
S1AP provides the signaling service between E-UTRAN (eNB) and the EPC (MME). S1AP services are divided into two groups:

- Non-UE-associated services: They are related to the whole S1 interface between the eNB and MME utilizing a generic signaling connection.
- UE-associated services: They are related to one UE. S1AP functions that provide these services are associated with a UE-associated signaling connection that is maintained for the UE.

For further details, please refer to 3GPP TS 36.413 [17].

eNodeB–eNodeB
Figure 1.8 shows the control plane protocol stack between two eNB(s) (X2 control reference point).

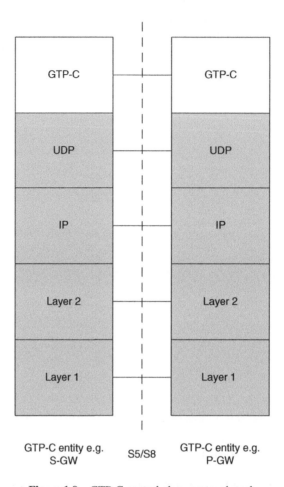

Figure 1.9   GTP-C control plane protocol stack

X2-Application Protocol (X2AP)
X2AP is used for mobility management procedures between two eNB(s) including handover preparation. In general, it is used to maintain the relationship between two eNBs. For further details, please refer to 3GPP TS 36.423 [18].

MME–S-GW–P-GW
Figure 1.9 illustrates the control plane protocol stack between any two network elements that support the GTP-C protocol. This can apply for MME–S-GW (S11 interface), MME–MME (S10 interface), and S-GW–P-GW (S5/S8-GTP-C-based interfaces).
   Short description for GTP-C (3GPP defined protocol) is provided here. Other protocols are based on standard IETF techniques.

GPRS Tunneling Protocol for the Control Plane (GTP-C)
Main purpose of GTP-C is to establish and maintain bearer tunnels (GTP-U tunnels) associated with bearer contexts in LTE/SAE for user sessions that require a certain QoS. For that purpose,

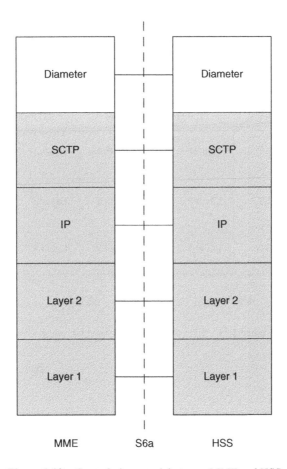

Figure 1.10   Control plane stack between MME and HSS

GTP-C enables the exchange of tunnel identifiers and tunnel addresses (IP addresses, port numbers) between the two involved entities. Such tunnels exist between radio and core network and within the core network. In EPC, the GTP tunnel is established between eNB and S-GW and between S-GW and P-GW. Together with the bearer on the radio interface between UE and eNB, this results in an end-to-end bearer with a certain QoS spanning from the UE toward the P-GW. Besides that GTP-C is also used to transport mobility management messages in case of relocation/handover. For further details refer to 3GPP TS 29.274 [19].

MME – HSS
Figure 1.10 shows the control plane protocol stack between MME and HSS.

S6a DIAMETER Application
Main function of S6a is to support transferring of subscription and authentication data for user authentication and authorization between MME and HSS. DIAMETER is defined in RFC 3588 [20] and the S6a DIAMETER application in 3GPP TS 29.272 [21].

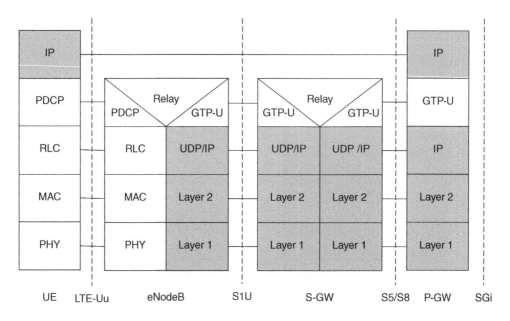

Figure 1.11   User plane stack between UE and P-GW

1.6.8.2   User Plane

Figure 1.11 shows the end-to-end user plane stack for a UE connecting toward a P-GW in EPS. The protocol stack is similar to the control plane stack. The only new protocol being introduced here is GTP-U.

GPRS Tunneling Protocol for the User Plane (GTP-U)
This protocol tunnels user data between eNodeB and eNodeB, eNodeB and S-GW, as well as between S-GW and P-GW. GTP-U encapsulates all user plane packets. For further details on GTP-U, please refer to 3GPP TS 29.281 [22].

1.6.8.3   Summary of Reference Points and Protocols

Table 1.3 summarizes the reference points and protocols used for LTE/SAE and PCC.

1.6.9   Mobility Management

In a broader sense "mobility management" includes all procedures used to support registration and mobility of a UE, such as the following:

- Selecting a network.
- Attaching to and detaching from the network.

Table 1.3 Reference points and protocols for LTE/SAE and PCC

Reference point	Protocols	Specifications
LTE-Uu	CP:PHY/MAC/RLC/PDCP/RRC	TS 36.300 [16]
	UP:PHY/MAC/RLC/PDCP	
UE – MME	NAS (EMM, ESM)	TS 24.301 [14]
X2	CP: X2AP	TS 36.423 [18]
	UP: GTP-U	TS 29.274 [19]
S1-MME	S1-AP	TS 36.413 [17]
S1-U	GTPv1-U	TS 29.281 [22]
S5	GTPv2-C/GTPv1-U	TS 29.274 [19]/TS 29.281 [22]
	PMIPv6	TS 29.275 [23]
S8	GTPv2-C/GTPv1-U	TS 29.274 [19]/TS 29.281 [22]
	PMIPv6	TS 29.275 [23]
S6a	DIAMETER	TS 29.272[21]
S9	DIAMETER	TS 29.215 [24]
S11	GTPv2-C	TS 29.274 [19]
Sp	Not specified in 3GPP	Not specified in 3GPP
Gx	DIAMETER	TS 29.212 [25]
Gxa	DIAMETER	TS 29.212 [25]
Gxb	Not specified in 3GPP	Not specified in 3GPP
Gxc	DIAMETER	TS 29.212 [25]
SGi	IPv4, IPv6, RADIUS, DIAMETER, DHCP	TS 29.061 [10]
Rx	DIAMETER	TS 29.214 [12]

- Maintaining the connection to the network while the UE is moving (referred to as "handover").
- Keeping the network informed about the present location of the UE in order to be reachable for paging even after the connection to the network has been released.
- Re-establishing the connection between UE and network when the UE needs to send uplink signaling or user data or when the UE was paged by the network because the network wants to send downlink signaling or user data.

Furthermore, certain security-related tasks are usually performed as part of the mobility management procedures: authentication, confidentiality protection of subscriber's identity, and confidentiality and/or integrity protection of signaling messages and user data.

Readers interested in detailed call flows for Attach, Detach, TA Update, Paging, Service Request, and Handover procedures in addition to the overview provided in this section may refer to the Appendix.

1.6.9.1 Attach

Each UE needs to register with the network to receive EPS services. This registration is called "network attachment." Always-on IP connectivity for a UE is enabled in EPS by establishing a so-called default EPS bearer during network attachment. The UE may request an IP address during the Attach procedure. The Attach procedure is triggered by the UE by sending an Attach Request message to the network. This message terminates at the MME. The Attach Request message is encapsulated in a RRC message toward the eNB and in a S1-MME control message to the MME. The MME that terminates the Attach Request message is called "serving MME." Its address is stored in the HSS.

One purpose of the Attach procedure is to authenticate the UE and activate integrity protection and ciphering of NAS messages exchanged between UE and MME. Necessary information for the authentication of a particular UE is obtained from the HSS. This information consists of so-called authentication vectors that are generated in the HSS from the keys stored in the subscription data.

If dynamic policy control is applied, the P-GW obtains PCC rules for the UE from the PCRF during the Attach procedure. If dynamic policy control is not deployed, the P-GW may apply local QoS policies. This could result in the establishment of a number of dedicated bearers for the UE in association or combination with the default bearer.

Finally, the UE is registered with the network to receive EPS services. It can request for specific resources to run a certain application and perform handover when radio coverage conditions change.

1.6.9.2   Detach

The Detach procedure allows the UE to inform the network that it does not want to access the network any longer and allows the network to inform the UE that it does not have access to the network any longer. Detaching a UE implies that all PDN connections and associated bearers are released.

The UE is detached either explicitly or implicitly. In case of an explicit detach, the network or UE explicitly requests detach and signal with each other. In case of an implicit detach, the network detaches the UE without notifying the UE. This is typically done when the network presumes that it is not able to communicate with the UE, for example, due to lack of radio coverage.

1.6.9.3   Tracking Area Update

Once the UE is successfully attached, it needs to keep the network informed about its current location in order to be reachable for downlink signaling and user data (incoming voice call or SMS) even when the UE is in "idle mode," that is, when the signaling connection between UE and network has been released. Moving in idle mode and keeping the network informed about its current location is in general referred to as "idle mode mobility." Although normally not active in idle mode the UE periodically wakes up and monitors the broadcast channel in order to receive information about incoming signaling or data traffic.

E-UTRAN cells are combined to Tracking Areas. A UE camping on an E-UTRAN cell in idle mode is listening to the system information broadcast in this cell, which includes the identity of the TA the cell belongs to. When the UE moves to another cell and the received TA identity indicates that the new cell belongs to a TA to which the UE is currently not registered (i.e., the TA is not in the list of registered TAs stored in the UE), the UE initiates a TA updating procedure.

In the simplest case, if the new cell and the new TA are served by the same MME to which the UE is already registered, only two NAS messages, TA Update Request and TA Update Accept, need to be exchanged between UE and MME. Otherwise, if the new cell is served by a new MME, or if the MME decides to change the S-GW, further network entities (HSS, old MME or old SGSN, old S-GW, P-GW) need to be involved in the procedure.

The TA or list of TAs allocated by the MME during the TA updating procedure or attach procedure can be used by the MME to page the UE in a certain area, when the signaling connection to the UE has been released and the network needs to send downlink signaling or user data. The MME can page the UE first in the last known TA, if the UE does not respond, paging can be performed in a wider area (e.g., in neighboring TAs) before the MME pages the UE in all TAs of the allocated TA list. The MME knows the location of a UE only up to the granularity of a TA, and it has no knowledge in which particular cell of a TA the UE is currently camping on.

1.6.9.4 Paging

The paging procedure is normally initiated by the network to request the establishment of a NAS signaling connection toward an UE that is in idle mode. In addition, the network can also initiate paging to inform the UE in RRC-IDLE or RRC-CONNECTED mode about system information change, inform the UE about an impending warning notification (e.g., for CMAS and ETWS as described in chapter 2). It can also be initiated by the network to request the UE to perform re-attach when the network has lost UE context due to a failure situation.

The trigger for paging a UE is usually a mobile-terminated transaction destined for the UE such as downlink data, an incoming SMS, or voice call.

It is up to the network (i.e., the MME) to decide how and with which priority to page the UE. The paging could be, for example, started in the last known cell and then extended to a wider area such as TA and TA list. Priority of paging can be determined by the bearer QCI and ARP.

When the UE is in RRC-IDLE state, the UE periodically monitors for a paging message from the network. The monitoring frequency in the UE is determined by the "Discontinuous Reception (DRX) cycle in idle mode." This idle mode DRX value can be configured via the Broadcast Control Channel (BCCH) and the NAS layer. If the UE has to monitor paging periodically, this consumes some power (i.e., battery life) in the UE. Thus, in Release-12, 3GPP introduced a new mode called "Power Saving Mode (PSM)" in order to provide additional means to save battery life for devices that normally require only infrequently mobile-originated transactions. When UE is in PSM, idle mode procedure does not apply, that is, UE does not listen to paging nor perform measurements. Thus, the UE is not reachable for mobile-terminated transactions. This is meant mainly for devices (e.g., with frequency such as twice a day transmission, 8 bytes a day transmission) for which mobile-terminated transactions are very infrequent or some delay in mobile-terminated transactions is acceptable without impacting the end user experience.

1.6.9.5   Service Request

The purpose of the Service Request (SR) procedure is to move the UE from ECM-IDLE to ECM-CONNECTED state and establish EPS bearers when user or signaling data have to be transmitted. Other purposes are to trigger MO/MT CS Fallback (CSFB), explained later in this chapter, and Proximity Services procedures (see Chapter 4).

If the UE has a pending mobile-originated transaction when it is camping in E-UTRAN, then the UE initiates a Service Request procedure toward the network. If the network has a pending mobile-terminated transaction, the network initiates a paging procedure first. Once the paging message has been successfully processed by the UE, it responds with a Service Request message toward the network to establish the necessary bearers.

The NAS Service Request message is used for fast re-establishment of the NAS signaling connection and the user plane bearers. In order to meet the strict performance requirements for EPS when it is sent as a response to a paging message, the NAS Service Request message transmitted over the air was carefully designed to fit within a single-radio transport block so that the message need not be segmented over the air interface.

When new features caused the addition of information elements, this requirement could no longer be met. For this reason, a new Extended Service Request (ESR) message was defined. ESR is used in special cases (e.g., to trigger a CS Fallback procedure) when additional parameters need to be sent. ESR follows the layout of regular EMM messages.

Thus, the Service Request procedure can be initiated using a regular Service Request or an Extended Service Request. The detailed conditions for when to use Service Request or Extended Service Request are specified in 3GPP TS 24.301 [14].

Successful completion of a Service Request procedure is determined by the UE either based on indication of successful establishment of radio bearers (when the UE continues to remain in E-UTRAN) or based on successful intersystem change (in case of CS Fallback).

1.6.10   Intra E-UTRAN Handover

1.6.10.1   General

"Handover" refers to situations where the UE is moving from one cell to another while it maintains a signaling connection with the network. In case of intra E-UTRAN handover, the UE moves from one LTE cell to another one. Inter-RAT handover refers to the case where source and target cell belong to different radio technologies, for example, UE moves from an E-UTRAN to a UTRAN cell.

In case of intra E-UTRAN handover, the UE can be handed over from the currently serving eNB ("source eNB") to a new one ("target eNB"). This handover procedure considers also existing data connections, that is, data connections are moved from the source to the target eNB. The handover process can even lead to a change of the serving S-GW and/or MME once the UE has moved into a service area that is no longer served by the old S-GW and/or MME. Only the P-GW as mobility anchor point is never changed during handover.

Two different signaling procedures have been defined for handover: X2-based and S1-based handover, named after the interface used for the exchange of signaling messages during the handover preparation.

1.6.10.2   X2-Based Handover

If both eNBs are connected to the same MME, source eNB and target eNB can exchange the S1AP signaling for the handover preparation and execution directly via the X2 interface (for details, see 3GPP TS 36.300 [16]).

The MME is involved in the procedure only during the handover completion phase when the UE has already been handed over to the target eNodeB. During the handover, the target eNB sends an S1AP message (Path Switch Request) to the MME indicating that the S1 interface needs to be switched from the source to the target eNB. Through this message MME is also aware that the UE has moved into a new cell. The MME updates the S-GW with the new address information for downlink user data and confirms the relocation of the S1 interface toward the target eNodeB. If the MME wants to use a different S-GW, it can allocate resources on the new S-GW and provide the target eNB with the necessary address information for sending uplink user data.

During the handover execution, downlink packets are forwarded by the source eNodeB to the target eNodeB. Once the S-GW receives the command to switch the user plane to the target eNodeB, it sends one or several "end marker" packets toward the source eNodeB to assist the target eNodeB with the reordering of packets.

After completion of the handover, the UE receives the system information broadcast in the target cell. If the target cell belongs to a TA to which the UE is not registered, the UE has to initiate a TA update procedure.

1.6.10.3   S1-Based Handover

S1-based handover is used in all cases when X2-based handover cannot be used, for example, when source eNB and target eNB are served by different MMEs (indicated by the TA identity of the target cell). During the handover preparation, the signaling information is exchanged between source eNB and target eNB via the involved MME(s), that is, messages are sent via the S1 interfaces, and possibly via the S10 interface between the MMEs. Independent of a possible MME change also the S-GW can be changed during S1-based handover.

In case of MME change, the source MME transfers context information about the UE to the target MME, including the EPS security context, and mobility management and session management information. For proper routing of uplink data packets, the MME provides the target eNB with the S-GW address.

During the handover execution, downlink packets are forwarded by the source eNB to the target eNB either directly or indirectly, that is, via the target S-GW.

After completion of the handover, the UE receives the system information broadcast in the target cell. If the target cell of the handover belongs to a TA to which the UE is not registered, the UE initiates a TA updating procedure.

1.6.11   Security

Communication via the radio interface is vulnerable to various security attacks such as eavesdropping, "man-in-the-middle" attacks, or subscriber tracking.

To protect the subscriber against eavesdropping, EPS supports encryption of signaling and user data.

In a "man-in-the-middle" attack, a "false base station" impersonates a "real" base station toward the UE and a UE toward the network, and relays, possibly modifies, the signaling between UE and network. To avoid "man-in-the-middle" attacks, the EPS supports integrity protection of signaling data.

The use of ciphering in a network is optional for the operator, whereas the use of integrity protection is mandatory. Without activation of integrity protection, the UE will not be able to successfully attach to the network and receive services. As an exception to this rule, the network can activate a so-called null integrity protection algorithm (see 3GPP TS 33.401 [26]) during an emergency attach for an unauthenticated, (U)SIM-less, emergency call. The null integrity protection algorithm is effectively providing no protection.

Before ciphering and integrity protection can be activated, UE and network need to establish an EPS security context. This security context is either created during an authentication procedure or derived from a UMTS security context during intersystem change from GERAN/UTRAN to E-UTRAN.

Once the EPS security context is used by the MME by means of a security mode control procedure, the UE will send all NAS messages integrity protected, including the initial NAS messages for subsequent network access and for re-attaching when the UE has temporarily detached from the network. Integrity protection in downlink direction and ciphering of NAS messages is started by the MME after successful authentication of the UE or when the network has verified the integrity protection of NAS messages sent by the UE by means of an already available EPS security context. Each time the UE accesses the network and establishes a new signaling connection, the MME needs to restart integrity protection and ciphering of NAS messages.

Security on AS level, including the ciphering of user data, is controlled separately by AS signaling procedures. Each time the UE accesses the network and establishes a new signaling connection, the MME needs to restart AS security.

A further aspect of subscriber confidentiality is the protection against location tracking of subscribers by third parties. In certain situations, signaling messages containing a subscriber identity need to be sent unciphered. To protect the subscriber against tracking, UE and network are using a temporary user identity whenever possible. During the initial attach procedure, the UE may need to identify itself toward the network with its permanent subscriber identity, the IMSI, but for subsequent accesses the UE will use the Globally Unique Temporary Identity (GUTI), which is assigned by the MME during attach procedure or the UE will use the S-TMSI which is a part of the GUTI. The GUTI is regularly reallocated by the MME via security-protected signaling.

In the downlink direction, the network is normally using the S-TMSI to page the subscriber.

An additional permanent identity related to the UE, the International Mobile Equipment Identity (IMEI), can be retrieved by the MME only via integrity-protected signaling. Thus, the IMEI can only be requested by an authorized network entity and is usually transmitted in ciphered form.

In addition to security on a per user basis, also security associations for links such as S1 (eNB to MME and S-GW) and X2 (eNB to eNB) providing integrity and confidentiality protection for all control and user plane traffic going via these interfaces exist. Figure 1.12 provides an overview of the overall EPS security architecture.

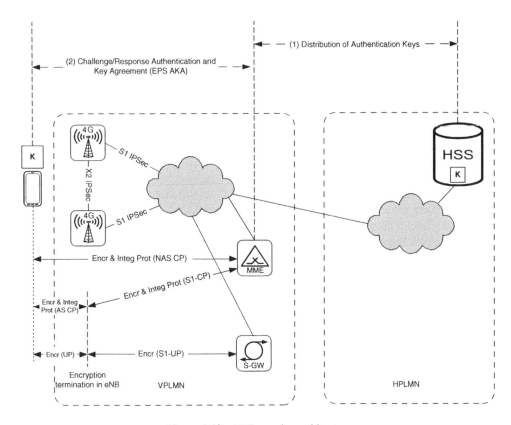

Figure 1.12   EPS security architecture

Following are the list of per-user security associations.

1. RRC signaling between UE and eNB is integrity and confidentiality protected.
2. User plane between UE and eNB is confidentiality protected.
3. NAS signaling between UE and MME is integrity and confidentiality protected.

Following are the list of security associations that are user independent.

1. S1-C (signaling) between eNB and MME is integrity and confidentiality protected.
2. S1-U (user data) between eNB and S-GW is integrity and confidentiality protected.
3. X2 (signaling and user data) between two eNBs is integrity and confidentiality protected.
4. Various signaling protocols for core network internal interfaces are integrity and confidentiality protected.

For more information related to security aspects, we refer the reader to 3GPP TS 33.401 [26]
and 3GPP TS 33.402 [27].

1.6.12   Charging

Logical charging functions in the EPC network are responsible to collect data for charging and billing purposes. These charging functions are specified in 3GPP TS 32.240 [28], together with the reference points that are used to transfer charging events and consolidate charging information between those functions. The mapping of the logical charging functions to the EPC architecture is described in 3GPP TS 32.251 [29].

1.6.12.1   Charging Principles

EPC nodes provide functions that implement Online Charging and/or Offline Charging mechanisms on bearer level, in particular for the PDP context/IP-CAN bearer, and on Service Data Flow (SDF) level. Online Charging performs traffic supervision in real time, while Offline Charging records the usage of resources (e.g., number of sent/received packages or volume or the duration of voice calls). The collected resource usage data can be used for billing of individual subscribers, inter-operator accounting, or general statistical purposes.

Offline Charging is a process where charging information for network resource usage is collected concurrently. The information is passed from the Charging Trigger Function (CTF) to the billing system. Offline charging does not affect the rendered service in real time.

Online Charging, on the other hand, is a process where charging information for network resource usage is not only collected concurrently during their usage, but also the authorization for using these network resources must be obtained before the actual usage. This authorization is granted by the Online Charging System (OCS) upon request from network functions such as GGSN or P-GW/PCEF. More details of the OCS architecture can be found in 3GPP TS 32.296 [30].

Offline and Online Charging can be performed simultaneously and independently for the same chargeable event, for example, a voice call or packet data transfer.

1.6.12.2   EPC Charging

In case of EPC charging, the following functional entities provide charging data:

- The SGSN, S-GW, and ePDG (see chapter 1.10) record user's access to VPLMN resources. In addition, the SGSN records user's mobility management activities, SMS, and Multimedia Broadcast/Multicast Service (MBMS) usage.
- The P-GW and GGSN record user's access to external networks such as the Internet.
- The MME record SMS usage when the "SMS in MME" feature is supported.

The functional entities in EPC differ with regards to the charging information they are able to supervise and report. So-called Charging Characteristic profiles stored in the HSS specify under which conditions a certain charging event is generated. Three default profiles exist: one for the non-roaming case, one for the local breakout roaming case, and one for the home routed roaming case. Besides these profiles stored in HSS, SGSN/S-GW and GGSN/P-GW/TDF can also use locally configured profiles.

1.6.12.3 Offline Charging

If a subscriber is using EPS network resources, the corresponding charging information is collected by the EPC functions serving the subscriber. SGSN and S-GW capture information regarding the usage of radio network resources as well as data pertaining to mobility management, while GGSN, ePDG, and P-GW collect charging information that relate to the utilization of external data network resources. The subscriber's usage of resources in the EPC network itself is recorded by each of them. Access to EPC network resources is provided by PDP contexts/IP-CAN bearers that offer the user a logical connection to services with a certain QoS. An APN identifies the service to which access is provided and network resources are consumed. Examples of such services are IMS or Internet access.

The EPC entities collect information pertaining to these PDP context/IP-CAN bearers and the resource utilization that comes along with their usage. EPC charging comprises collecting information about transferred data volume separated for downlink and uplink traffic and categorized by the provided QoS and used protocols, the duration of this usage (i.e., how long the PDP context/IP-CAN bearer is activated), destination, and source address, the APN as well as the location of the UE, that is, the network to which the UE is currently attached to. Regarding the GGSN and P-GW, the accuracy of this location information is limited to the SGSN address, whereas in SGSN and S-GW also the E-UTRAN cell identity is available. The MME plays a role for charging only in case of SMS offline charging when "SMS in MME" feature is supported.

User's activities are recorded in charging events by the Charging Data Function (CDF) and finally formatted into so-called Charging Data Records (CDR) generated in the entities serving the user. The CDR is transferred to a Charging Gateway Function (CGF) for further processing and from there to the Billing Domain (BD).

For EPC charging the creation of a CDR is triggered by a charging event during PDP context/IP-CAN bearer activation (e.g., in P-GW). Thus, EPS bearer context activation is, for example, a chargeable event. At the same time, volume counters for this context are initialized counting the transferred data volume in uplink and downlink. Upon occurrence of certain chargeable events, such as change of QoS or radio access type, these volume counters are captured together with a timestamp and the applied QoS in the collecting node. Other charging triggers, such as deactivation of the PDP context/IP-CAN bearer or operator-defined limits for time or volume lead finally to the closure of the CDR. If the IP-CAN bearer remains active, a new charging record is created. Besides data volume or elapsed time, other data such as user's IP address, protocol type, and APN are also stored in a CDR.

Common information to all CDRs is the IMSI, optionally the MSISDN and the IMEI, identifying the subscriber and the UE.

The CGF receives charging records over the Ga reference point, if it is not integrated in the node sending the records. The CGF can also be a separate entity or part of the BD . The BD is responsible to create the final bill toward the subscriber (e.g., on a monthly basis). In all these cases, charging records are transferred from CDF to CGF via the GTP protocol. The transfer of the CDR files from the CGF to the BD uses the Bp reference point.

1.6.12.4 Online Charging

In contrast to offline charging, the solutions for online charging on the SGSN/S-GW and the GGSN/ePDG/P-GW differ significantly.

The SGSN uses legacy non-IP techniques for online charging while EPC online charging at the GGSN, ePDG, and P-GW uses DIAMETER. Online charging may apply on a PDP context/IP-CAN bearer level or on individual service flow level. EPC online charging at the P-GW and TDF is utilizing the Ro interface and the associated DIAMETER Credit-Control application toward the OCS.

The P-GW (i.e., the PCEF as part of the P-GW) collects charging information on the IP-CAN bearer level for each user separated for uplink and downlink. The defined chargeable events correspond to the ones for offline charging for debiting, that is, start and stop of a PDP context/IP-CAN bearer, reaching of time or volume limits, as well as QoS or tariff time changes. If such events occur, the OCS is informed by the P-GW/PCEF and need to authorize the event beforehand, for example, that a new bearer can be established.

Upon establishment of a PDP context/IP-CAN bearer the P-GW/PCEF requests authorization of this event from the OCS. The OCS either authorizes the request when the answer contains a certain volume and/or time quota or denies the context establishment. If authorization is confirmed, a volume counter and/or time measurement is granted in the P-GW/PCEF that debits the counters based on the traffic transmitted via the PDP context/IP-CAN bearer.

If for an established PDP context/IP-CAN bearer the supplied quota is used, the OCS is requested to perform a re-authorization. In this case the used volume count is reported by the P-GW/PCEF to the OCS. When a change of charging conditions has occurred, this information is used to determine how much of the quota has been consumed for the old QoS and during the tariff time.

Generally, the OCS replies to the P-GW/PCEF with quota and instructions how the P-GW/PCEF shall further proceed, for example, continue or terminate an ongoing session.

1.6.12.5   Flow-Based Bearer Charging

To allow for a service-based charging below bearer level, Flow-Based Charging (FBC) has been introduced with 3GPP Release 6 (see 3GPP TS 23.203 [4]). It adds new functionalities to the EPC network and especially extends the capabilities of the P-GW in such a way, that it is now able to identify different Service Data Flows (SDF) within a single PDP context/IP-CAN bearer.

From the point of view of charging information collection, FBC can be considered as an extension of the classical EPC charging by being able to sub-categorize the total data volume of a PDP context/IP-CAN bearer by the different flows the context/bearer contains. For each of these SDFs, an own uplink and downlink volume counter is required.

With FBC it is possible to employ the same charging models for both offline and online charging and irrespective of whether the subscriber is prepaid or postpaid. FBC provides a high degree of flexibility regarding which SDF to consider and how to recognize a SDF by means of charging rules.

The overall FBC concept is realized by three functional elements: the Policy and Charging Rules Function (PCRF), the Application Function (AF), and the Policy and Charging Enforcement Function (PCEF). The overall PCC architecture is described in Section 1.6.7.

The PCRF provides Charging Rules via the Gx reference point to the PCEF and decides which rules have to be applied based on data received from the PCEF, such as user- and bearer-related information, as well as from the AF, which provides session- and media-related information. Such rules contain information how packets belonging to a flow can be identified and how they shall be treated, in particular regarding the applicable QoS. Charging rules need to be specified for each particular service.

The AF provides services to the user for which PDP context/IP-CAN bearer resources are required. The AF can provide additional information to the PCRF for charging rule selection or generation, such as an application identifier, specific user information, and packet filters allowing for the identification of packets belonging to the respective service data flows.

The PCEF is responsible for identifying the user data traffic based on the charging rules received from the PCRF. It is used for both online and offline charging. In case of online charging, the PCEF has also to take care of Credit Control, that is, maintaining the assigned quota as well as communication with the OCS. The PCEF can be located in the P-GW or in case of untrusted WLAN access in the ePDG.

In case of offline charging, charging information collected by the PCEF is written into a PGW-CDR. When flow-based offline charging is activated in a P-GW, classical IP-CAN bearer charging is not done anymore.

In case of online charging, the PCEF creates charging events to request authorization regarding the resource usage being caused by a SDF. On IP-CAN bearer activation a FBC Credit-Control session is started. The OCS replies with a charging event response either granting an initial quota or denying the service request. Following requests toward the OCS may be triggered when the subscriber starts using a new service, that is, when the PCEF encounters one or more new SDFs or when the granted quota is used up. The charging session is closed when the IP-CAN bearer is deactivated or the OCS indicates session termination as a result of the fact that the subscriber has consumed his or her credit.

1.6.12.6 Charging Rules

FBC uses charging rules to identify and charge SDFs. A SDF consists of one or more IP flows that shall be treated and charged as a whole.

A SDF is identified by means of SDF filters. Such a filter may be an IP 5-tuple containing of destination/source IP addresses, destination/source port numbers and protocol (TCP, UDP, SCTP), possibly with wildcards, and also other filters for application protocol or content recognition and form part of a charging rule. The wildcard represents the so-called default SDF, which comprises all traffic. If it is the only defined SDF in a PCEF, the resulting behavior of FBC corresponds to the classical EPC charging on bearer level. Otherwise, the default SDF captures all traffic that does not match at least one of the more specific flow filters.

Charging rules in the PCEF are applied by matching received packets with the SDF filters that are part of these charging rules. Pre-defined charging rules are completely configured, that is, they contain all necessary information to be applied. They may already be installed on the PCEF or provided by the PCRF via the Gx reference point when needed. In contrast, dynamic charging rules are newly generated or completed dynamically using application-specific criteria to identify the SDF. This information is provided by the AF to the PCRF on request over Rx.

Besides SDF filters, charging rules may comprise further information that, for instance, define how the charging process shall take place, that is, whether time- and/or volume-based resource usage information shall be provided, which precedence the charging rule has in case of a rule overlap, whether online or offline charging has to be applied, and identifiers for charging correlation and service identification.

Charging rule provision takes place over the Gx reference point. The provision of rules can either occur following a charging rule request by the PCEF (caused, e.g., by bearer establishment, bearer modification, or QoS change) or unsolicited by the PCRF as the result of new information received from an AF or via notification received from the OCS.

The Rx reference point plays an important role for dynamic charging rule generation and completion and is used for exchanging information between the PCRF and the AF. When the AF becomes aware of new media used by an application, it provides this information, which in particular relates to the different flows and how they can be identified, to the PCRF, which in turn generates or completes dynamic charging rules by adding SDF filters. This procedure may also be initiated by a PCEF requesting charging rules and a PCRF recognizing that specific information is required for dynamic rule provisioning. The PCRF then contacts the corresponding AF to acquire the needed information.

1.6.12.7 MBMS Charging

For MBMS charging, the BM-SC contains an integrated CTF that generates charging events for mobile subscribers receiving services through the MBMS user service and/or for content providers delivering content through the MBMS bearer service. Transactions involving the content provider are recorded per subscriber. Online charging at the BM-SC utilizes the Ro reference point while offline charging at the BM-SC utilizes the Rf reference point.

The MBMS GW collects charging information for each MBMS bearer service that is activated. Following information is reported by the MBMS GW:

- Start of MBMS bearer context: A CDR for the MBMS bearer context is created and the data volume is captured for the context.
- MBMS bearer context termination in the MBMS GW.
- Expiry of an operator configured time or data volume limit per MBMS bearer context. This event closes the MBMS bearer context CDR and a new CDR is opened, if the context is still active.
- Change of charging condition, for example, tariff time change. In this case the current volume count is captured and a new volume count is started.
- Expiry of an operator configured change of charging condition limit per MBMS bearer context closes the MBMS bearer context CDR. A new CDR is opened, if the MBMS bearer context is still active.

More details on MBMS charging can be found in 3GPP TS 32.273 [31].

1.7   IP Multimedia Subsystem

3GPP Rel-5 in 2001 was the first release where the basic architecture and procedures of the IP Multimedia Subsystem (IMS) were specified. While mobile operators were initially reluctant to commercially deploy IMS in their networks to replace existing voice and SMS, with the introduction of LTE as a pure packet-based system without a CS voice component and the general VoLTE profile specified by GSMA in IR.92, IMS is becoming more and more important for mobile operators. Early mobile deployments of IMS were mainly focused on non-voice

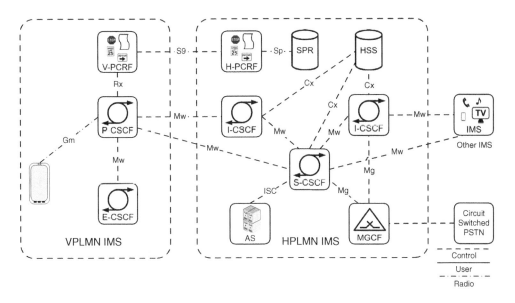

Figure 1.13   IMS roaming architecture

applications (messaging and presence) but nowadays (end of 2014) commercial mobile IMS deployments are planned or even rolled out introducing VoIP and other multimedia services on top of IMS.

In a nutshell IMS can be seen as a general purpose platform for the establishment, modification, and teardown of multimedia sessions between end points to exchange content such as voice, video, messages, and files with the necessary QoS. In principle, IMS can work on any packet-based network providing suitable data rates. One design goal was to provide more flexible communication service creation in an "Internet style" while, on the other hand, being able to handle the limitations of mobile communication systems. For that purpose, the Session Initiation Protocol (SIP) as specified by IETF RFC 3261 [32] was chosen to maintain sessions between UEs or between UEs and servers. The Session Description Protocol (SDP), specified in IETF RFC 4566 [33], transported within SIP message bodies allows end points to exchange information on multimedia content like the used codec. IMS works on the application layer and uses EPS capabilities to setup certain IP bearers with specific QoS between the IMS-capable UE and P-GW. Interworking between IMS and traditional networks such as the 2G/3G CS domain or the fixed public telephone network is enabled by the Media Gateway Control Function/Media Gateway (MGCF/MGW) where the MGCF interworks between SIP and CS protocols like ISDN User Part (ISUP) and the MGW interworks between different transport planes (e.g., between different codecs).

The basic IMS architecture is shown in Figure 1.13.

The main components of IMS are the so-called Call Session Control Functions (CSCF) that can be seen as SIP proxies. Four types of CSCF exist: P-CSCF, S-CSCF, I-CSCF, and E-CSCF.

The Proxy CSCF (P-CSCF) is the outermost SIP entity toward the served UE.

In case of "IMS roaming" the P-CSCF is located in the visited network (VPLMN). To select a proper P-CSCF in the VPLMN, GSMA has specified the well-known "IMS APN" which is resolved by the visited MME into a local P-GW address. The local P-GW selects a P-CSCF in the VPLMN and provides the address to the UE (note that in general P-GW and P-CSCF are always located in the same network, either both in HPLMN or both in VPLMN). Otherwise, the P-CSCF is located in the home network. When a UE registers, it is assigned a P-CSCF as entry point toward the IMS network. The P-CSCF stores information related to that served UE while it remains registered and forwards any SIP message to or from the served UE on the Gm interface. The P-CSCF provides integrity and confidentiality for SIP messages, compression and generation of charging records. It provides also an interface to the PCC and Lawful Interception infrastructure. In PCC terms, the P-CSCF acts as an Application Function (AF) toward the PCRF. Registration in the IMS requires a special application on the UICC, the so-called ISIM (IMS SIM). The ISIM holds information about subscriber identities and keys used in IMS.

The Serving CSCF (S-CSCF) is the central IMS entity located in the home network of the IMS subscriber. It acts as SIP registrar for the UEs, that is, the UE registers its contact address (IP address) at the S-CSCF when it connects to IMS. The S-CSCF retrieves the IMS user profile from the HSS. The user profile is used to authenticate and authorize a user and to execute services the user is subscribed to. Services are usually not executed by the S-CSCF itself but via special Application Servers (AS) that are connected to the S-CSCF through the ISC interface. Examples of Application Servers are Telephony Application Server (TAS) for value-added voice services (e.g., call forwarding) and conferencing, messaging, and announcement servers. Besides routing and address translation, the S-CSCF generates also charging records and interconnects with LI entities.

The Interrogating CSCF (I-CSCF) is the first contact point for SIP messages at the network boundaries. When receiving a SIP message destined for a user the I-CSCF selects an S-CSCF capable to serve that user. The I-CSCF does not store user-related data and it does not stay in the path for subsequent SIP messages. The I-CSCF is used when a UE registers to the network and when a session setup destined for a served user is received in the home IMS.

The Emergency CSCF (E-CSCF) handles emergency calls. Its main purpose is to retrieve location information about the caller and to forward the emergency call to a Public Safety Answering Point (PSAP) or emergency center.

The HSS is the central database located in the home network of a user, containing subscription and location-related information. Examples of these data are user identities (which are, e.g., SIP URIs), security keys used for authentication and authorization on IMS level, address of the S-CSCF serving the user, and the user profile containing a list of subscribed services.

The Media Gateway Control Function (MGCF) converts SIP messages into messages used in circuit-switched networks such as the Public Switched Telephone Network (PSTN) or CS domain of a mobile network. In addition, the MGCF controls Media Gateways (MGW). A MGW converts between the transport layer in a CS network such as TDM (Time Division Multiplex) and RTP (Real Time Transport Protocol) as used in IMS. The MGCF is responsible to allocate and maintain resources on a selected MGW and to provide necessary information received in signaling messages from its peers to the MGW to allow for transport conversion.

1.7.1 Summary of Reference Points and Protocols

Table 1.4 summarizes the reference points and protocols used for IMS.

Table 1.4 Reference points and protocols for IMS

Reference point	Protocols	Specifications
Gm	SIP	TS 24.229 [34]
Mw	SIP	TS 24.229 [34]
Cx	DIAMETER	TS 29.229 [35]
Mg	SIP	TS 24.229 [34], TS 29.163 [36]
ISC	SIP	TS 24.229 [34]

1.8   Voice and SMS in LTE

1.8.1   Voice

LTE/SAE is a pure packet-based system without an in-built circuit-switched voice component such as GSM and UMTS. Thus, the preferred way to provide voice and other multimedia services in LTE/SAE is via the IMS. For a better worldwide acceptability and interoperability of IMS over LTE, GSMA defined an IMS profile for voice and SMS in their IR.92 recommendation. For example, GSMA IR.92 has specified a well-known APN, the so-called IMS APN, to enable IMS roaming. This profile is referred to as "Voice over LTE (VoLTE)."

The LTE/SAE network indicates to the UE that VoIP in LTE is supported; more precisely that the TA(s) the UE is currently camping on provide sufficient QoS and coverage for VoIP. Therefore, an UE that has been provisioned as IMS VoIP capable can start an IMS voice session at its current location based on the received network indication.

However, during the early stages of LTE roll out, actual LTE coverage is spotty. A UE can start an IMS voice session in LTE/SAE but need to continue the voice call when it moves out of LTE coverage. To allow continuation of voice calls in 2G/3G CS domains (usually providing country wide coverage), the so-called Single Radio Voice Call Continuity (SRVCC) feature was introduced (see 3GPP TS 23.216 [37]). It extends the voice services coverage area and provides a good voice experience in the early stages of LTE. SRVCC requires bearer level handover between EPC and Circuit-Switched Core and switching of access legs between IMS and CS domain. The non-voice components (e.g., video streaming, file transfer) can be moved from LTE to 2G/3G PS via general handover procedures. Although 2G/3G CS coverage may allow voice calls all over a country, it is beneficial to start voice and multimedia calls in LTE/SAE whenever possible and even return to LTE/SAE as soon as possible. This is because a broadband network such as LTE/SAE together with a flexible service platform like IMS at the application layer can provide much more value-added services to the end user than the legacy 2G/3G CS systems. Benefit for the operator is that the user can be charged for these value-added services in addition to the normal voice service.

Some LTE/SAE networks may not support IMS-based voice service. In this case, voice service can only be provided over the CS domain of the existing 2G/3G networks. In order to accomplish that, the CS Fallback (CSFB) feature was defined by 3GPP (see TS 23.272 [38]). CSFB allows the UE to switch from LTE to 2G/3G network in a controlled manner to setup a CS voice call. Data connections can be handed over to the 2G/3G PS domain or (if not possible because of lack of resources) suspended in LTE/SAE. After the voice call has ended, the UE returns to LTE and data connections are resumed.

1.8.2   Short Message Service

SMS can be provided in LTE natively over IP (called "SMS over IP") via IMS (see 3GPP TS
23.204 [39]). There is no special requirement being placed on LTE/SAE for this feature, it is
just used to provide IP connectivity. The precondition is that the UE must have successfully
performed an IMS registration, is supporting the "SMS over IP" feature, and has been configured to use the feature. If this is the case, a SMS is encoded by the network or UE in a special
SIP message and send to the peer.

For the scenario in which the LTE operator also runs a 2G or 3G network, SMS over
LTE/SAE reusing CSFB mechanisms was specified. In this case, both UE and EPC must
support the SMS-specific procedures as defined for CSFB. Unlike CSFB voice, SMS over
EPC does not require the UE to tune to 2G/3G radio for SMS. The EPC network provides a
kind of tunnel between the UE and the CS Core for SMS delivery. The SMS is forwarded
between MME and MSC. The MSC is connected to the legacy SMS infrastructure (Short
Message Service Center) in the usual way. The UE is attached in both the EPS and CS
networks for SMS delivery.

A SMS can also directly be delivered by the SMS infrastructure to the MME and vice versa
via a DIAMETER-based interface, bypassing the MSC. This was introduced to support SMS
delivery to PS only UEs. PS only means that the UE does not have any CS subscription data in
the HSS. This feature is known as "SMS in MME." SMS in MME was introduced mainly to
address requirements from operators who do not deploy a CS core. SMS over IP (i.e., SMS over
IMS) could be one solution for these operators. However, SMS over IP requires an IMS/SIP
client in the UE, which is too heavy-weight for some types of devices such as smart meters or
dongles. Furthermore, inbound roamers whose home operators do not support IMS cannot be
offered SMS over IP in the VPLMN.

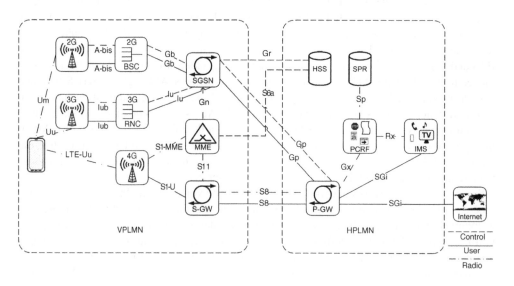

Figure 1.14   Interworking with Gn/Gp-SGSN

1.9   Interworking with 2G/3G Networks

1.9.1   Overview

3GPP has specified one way of interworking between LTE/SAE and existing 2G/3G networks by upgrading 2G/3G SGSN nodes to so-called S4-SGSN nodes. A S4-SGSN acts toward the EPC like a MME, that is, it has control plane interfaces to MME (S3 interface) and S-GW (S4 interface) and user plane interfaces to S-GW (S4), while connecting to 2G/3G radio access networks via the existing GPRS and UMTS interfaces. The S4 interface with the S-GW is used to manage bearers and possibly forward user plane traffic.

1.9.2   Interworking with Legacy Networks

If a mobile operator does not want to upgrade an existing SGSN (the so-called Gn/Gp-SGSN named after the interface between SGSN and GGSN in 2G/3G networks, see 3GPP TS 29.060 [8]) to a S4-SGSN, it is still possible to interwork these legacy SGSNs with the EPC. Such scenarios are important for a smooth introduction of LTE/SAE while leaving existing network elements untouched. The basic idea is to provide Gn/Gp interfaces at MME and P-GW, that is, from Gn/Gp-SGSN point of view the MME behaves like an SGSN and the P-GW behaves like a GGSN. Consequently, MME and P-GW must implement the protocols supported by Gn/Gp-SGSN, and additionally some changes in the mobility and session management procedures are required. It should be noted that this kind of interworking is only possible if GTP-based S5/S8 interfaces are used, as for PMIP-based S5/S8 handover between Gn/Gp-SGSN and MME/S-GW is not supported.

Figure 1.14 shows the Gn/Gp interworking architecture in the roaming case. The user plane is running from GERAN/UTRAN via the Gn/Gp-SGSN in the VPLMN to the P-GW in the HPLMN. The legacy Gr interface toward the HSS is MAP (Mobile Application Part) based.

The non-roaming architecture is similar to the roaming architecture with the difference that S5 is used between S-GW and P-GW, and Gn between Gn/Gp-SGSN and P-GW. If the Gn/Gp-SGSN supports Direct Tunnel (as defined from 3GPP Release 7 onwards) a direct user plane connection between UTRAN and P-GW is possible.

1.9.3   Functional Description

1.9.3.1   UE Aspects

In order to interwork with legacy 3GPP access technologies, an UE needs to be multiradio capable. It needs to support idle mode and connected mode mobility between E-UTRAN, UTRAN, and GERAN. There are two modes of interworking defined: single-radio operation and dual-radio operation.

In case of single-radio operation, the network controls the usage of radio transmitter and receiver in the UE in a way such that only one radio is operating at any time. This is optimized interworking and allows UE implementations where only one pair of physical radio transmitter and receiver is implemented. With dual-radio operation, multiple radio transmitters and receivers are operating simultaneously. Single-radio operation is an important mode because different access networks operate in different frequencies that might be close to each

Table 1.5 Reference points and protocols for interworking with 2G/3G networks

Reference point	Protocols	Specifications
Gn	GTP (v0 and v1)	TS 29.060 [8]
Gp	GTP(v0 and v1)	TS 29.060 [8]
S4	GTPv2-C	TS 29.274 [19]

other. So, dual-radio operation can cause high interference within the UE. Furthermore, it can consume additional power and reduce the overall performance and is more costly in terms of implementation.

1.9.3.2 E-UTRAN Aspects

The main additional function of the eNB is the support of mobility to and from UTRAN and GERAN. From eNB perspective, it needs to provide similar functionality for mobility to UTRAN and GERAN, for example, neighboring GERAN and UTRAN cells from the same network need to be configured in the eNB. Handover to and from UTRAN/GERAN is performed via the MME.

1.9.3.3 EPC Aspects

The S-GW acts as a mobility anchor for all 3GPP access systems. It functions as a GGSN toward SGSN. Although GGSN functions are mainly performed by the P-GW, this is not visible to the SGSN. The S-GW is controlled by the MME or SGSN, depending on the access network where the UE camps on (E-UTRAN or GERAN/UTRAN).

In order to support interworking, the MME needs to support signaling procedures with the SGSN. This is similar to the handover procedure supported by MME when MME relocation occurs. For interworking with legacy Gn/Gp-SGSN, the MME behaves like an SGSN.

1.9.3.4 Summary of Reference Points and Protocols

Table 1.5 summarizes the additional reference points and protocols used for interworking with 2G/3G networks.

1.10 Interworking with Non-3GPP Access Networks

Non-3GPP access networks refer to networks not using 3GPP access technologies, that is, not using GERAN, UTRAN, and E-UTRAN access. A typical and probably the most important example for a non-3GPP access network is a WLAN at a public hotspot, at a campus, or at home. Such a WLAN can, for example, use IEEE 802.11b/g/n radio technology. Interworking with non-3GPP access networks means providing access to the EPC and its services via a non-3GPP access technology and providing mobility between 3GPP and non-3GPP access (e.g., handover of connections from E-UTRAN to WLAN and vice versa). Interworking to non-3GPP access networks is described in detail in 3GPP TS 23.402 [3].

Basic principle of the non-3GPP access interworking architecture is that the P-GW is the IP mobility anchor point, that is, the P-GW is considered the "point of attachment" to external IP networks. In case of mobility between 3GPP and non-3GPP networks, the P-GW does not change.

From 3GPP perspective, non-3GPP access networks can be considered as either trusted or untrusted. It is operator's decision whether an access network is seen as trusted or untrusted. It does not depend on the access technology, but on operator policies and the business relationship between network operator and provider of a non-3GPP access network like a WLAN hotspot. The particular business relationship depends especially on the level of security provided by the access network and whether this is sufficient to allow access to the EPC. A non-3GPP access network can be seen as trusted for one operator while untrusted for another one.

Trusted non-3GPP access networks can be directly connected to the 3GPP core network. If an untrusted non-3GPP access network is used, the UE is connected to a kind of security gateway called Evolved Packet Data Gateway (ePDG) via an IPSec tunnel using the SWu reference point. The ePDG is located in the EPC.

Figure 1.15 gives an overview of the architecture with network-based mobility. The S2a interface connects mobile terminals to the core network over trusted non-3GPP access networks and the S2b interface is used for untrusted access networks. The functionality of S2a and S2b is quite similar and both can be implemented using either GTP or PMIP core network signaling. According to the specification, S2a can also be used in MIP Foreign Agent mode, however, it is not expected that this alternative will be widely deployed. When PMIP is used over S2a or S2b the trusted non-3GPP access network and the ePDG provide the Mobile Access Gateway (MAG) functionality required for PMIP, while the P-GW includes the Local Mobility Anchor (LMA) functionality. The interface between PCRF and ePDG is not specified. In roaming scenarios, that is, when the non-3GPP access network is connected to a VPLMN, the ePDG is in the VPLMN. In roaming scenarios, the 3GPP AAA server is located in the HPLMN and a 3GPP AAA proxy is located in the VPLMN.

Figure 1.16 shows the architecture when client-based (i.e., UE based) mobility based on Dual-Stack Mobile IP (DSMIP) is used. For simplicity not all interfaces are shown. Except for S2a, S2b, and S2c, interfaces in Figures 1.15 and 1.16 are the same. DSMIP between UE and LMA located in the P-GW runs over the S2c interface. In order to avoid user plane tunneling overhead over 3GPP access networks, the 3GPP access network is always the home link in terms of Mobile IP. Therefore, only S2c signaling is used over 3GPP access. The UE provides the DSMIP client function while the P-GW includes the DSMIP Home Agent function (LMA). In roaming scenarios, the 3GPP AAA server is again located in the HPLMN and a 3GPP AAA proxy is located in the VPLMN.

The basic non-3GPP interworking specification in 3GPP TS 23.402 [3] creates a general framework how to access EPC from a non-3GPP access network without putting any requirements on the WLAN. During the past couple of years, more and more 3GPP operators deployed WLANs and it has been recognized that without additional specifications these WLANs cannot be used as trusted non-3GPP access networks. The main technical issues are that in current WLANs the UE has no means to send handover and APN-related information to the network, and the support of multiple PDN connections is missing.

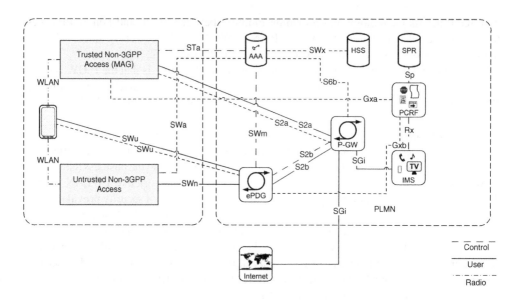

Figure 1.15 Non-3GPP interworking architecture with network-based mobility

In order to enable easy deployment of Trusted WLAN Access Networks (TWAN) in Release 11, a solution that can work without UE impacts was developed. This solution does not create any architecture level changes, just requires some enhancements to the S2a, STa, and SWx

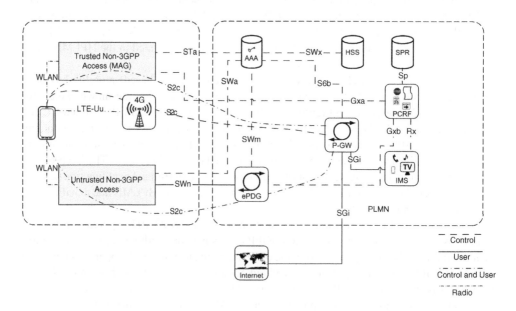

Figure 1.16 Non-3GPP interworking architecture with client-based mobility

interfaces. The AAA interfaces (STa and SWx) are enhanced to carry trusted WLAN authorization related parameters and additional subscriber data for trusted WLAN access. In addition, also GTPv2 support for Trusted WLAN Access over S2a was specified since it was recognized that for operators having only GTP-based interfaces, the deployment of GTP-based S2a might be easier than deploying PMIP-based S2a.

Owing to the lack of handover and APN indication, the solution in Release 11 does not support the following features:

– Handover between TWAN and 3GPP access with IP address preservation.
– Connectivity to a non-default APN (as it is not signaled by the UE).
– UE-initiated connectivity to additional PDNs.

Moreover, simultaneous access to EPC and non-seamless (without preserving UE's IP address) WLAN offload is not supported.

To overcome these limitations, 3GPP specified a new protocol between UE and an entity in the Trusted WLAN Access Network called Trusted WLAN Access Gateway (TWAG) in Release 12. The protocol is called WLAN Control Protocol (WLCP). WLCP signaling is transported over UDP/IP and enables management of PDN connectivity over TWAN. It provides the following functions:

– Establishment, termination, and handover of PDN connections.
– Request the release of a PDN connection by the UE or notify the UE of the connection release.
– IP address assignment.

1.10.1 Summary of Reference Points and Protocols

Table 1.6 summarizes the additional reference points and protocols used for interworking with non-3GPP access networks.

Table 1.6 Reference points and protocols for interworking with non-3GPP access networks

Reference point	Protocols	Specifications
SWa	DIAMETER	TS 29.273 [40]
SWm	DIAMETER	TS 29.273 [40]
SWu	IPSec/IKEv2	TS 24.302 [41]
SWw	WLCP	TS 24.244 [42]
	EAP	IETF RFC 3748 [43]
	EAP-AKA′	IETF RFC 5448 [44]
SWx	DIAMETER	TS 29.273 [40]
STa	DIAMETER	TS 29.273 [40]
S2a/S2b	GTPv2-C/GTPv1-U	TS 29.274 [19]/TS 29.281 [22]
	PMIPv6	TS 29.275 [23]
S2c	DSMIPv6	TS 24.303 [45]

1.11 Network Sharing

Various mechanisms exist for operators to share network deployment costs and increase country wide radio coverage. Increased coverage is an important use case in the beginning of Public Safety network rollouts. These mechanisms allow for sharing of equipment sites, radio network elements, spectrum/frequencies, and core network nodes.

The network sharing solution described in this section is a feature for a shared radio spectrum scenario (i.e., different operators share the same spectrum) where a single cell is broadcasting multiple PLMN ID(s). This feature was originally specified as an option for UMTS in Release 6 and was adopted for EPS in Release 8. In Releases 10 and 11 it was also specified for GERAN access. The feature is defined in 3GPP TS 23.251 [46] and is sometimes referred to as Multi Operator Core Network (MOCN) and/or Gateway Core Network (GWCN) depending on the core network configuration. In the MOCN configuration, operators share the radio access network (eNB) but operate separated core networks (MME, S-GW, P-GW, and HSS) while in the GWCN configuration also the MME is shared (but not S-GW, P-GW, and HSS). It is obvious that sharing RAN nodes provides much more benefits to the operators in terms of cost reduction than just sharing a few CN nodes.

In both configurations – MOCN and GWCN – the eNBs in the radio access network are shared in the same way and the UE behavior is also the same. While in the MOCN configuration the shared eNB connects to core networks of different operators, GWCN allows the MME to be shared and the MME connects to GW(s) and HSS of different operators. Sharing of an eNB or MME means basically that several operators (at least two) can use the same HW and SW resources but can configure parts of the shared node individually (depending on the functionality provided by the eNB/MME vendor). Figure 1.17 shows the MOCN and GWCN configurations.

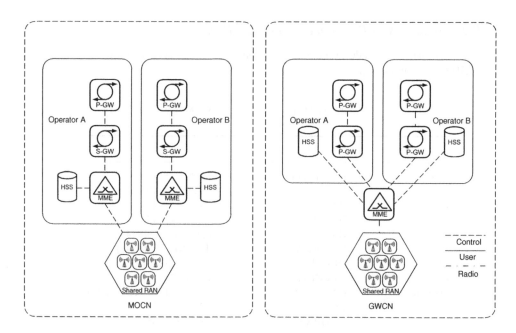

Figure 1.17 MOCN and GWCN configurations

1.11.1 UE-Based Network Selection

Each LTE cell in a shared area broadcasts multiple PLMN Identities (in maximum five) in a list where the first listed identity is the so-called primary PLMN. The broadcasted TA Code is common to all PLMNs. A UE decodes the broadcast system information and takes the information concerning all available core network operators into account in network and cell (re-)selection procedures.

When a UE performs an initial access to a shared network it selects one of the advertised networks (usually its home PLMN) and indicates the selected PLMN identity to the eNB.

1.11.2 RAN-Based Network Selection

The UE informs the eNB about the network identity of the chosen core network. On the basis of this information the eNB routes UE's initial access request to one of the selected operator's MMEs.

Once the UE gets admitted, the MME provides a temporary identity to the UE, which contains sufficient information to enable the eNB to direct subsequent messages to the same MME.

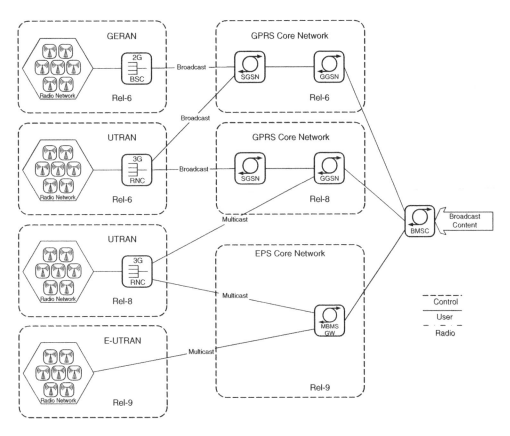

Figure 1.18 Evolution of MBMS user plane

1.12   Multimedia Broadcast Multicast Service

1.12.1   Principles

Multimedia Broadcast Multicast Service (MBMS) is a unidirectional point-to-multipoint service that allows simultaneous transmission of data from a single source (the content provider)
to a group of users located in a specific area. MBMS provides means to send data to a potential huge number of users in an efficient manner. For that purpose MBMS uses radio multicast
channels on the air interface and IP multicast techniques in the core network. Services that
may use MBMS transport capabilities can be divided into two types:

- Streaming services with a continuous data flow
- Download and play services.

Service examples include the following:

- Video distribution, Mobile TV, and Mobile Gaming via streaming or download
- Traffic announcements
- Content distribution such as downloading files, HTML pages, video, audio, or a combination
  of those and software updates to the device.

MBMS works in broadcast and multicast mode but only broadcast mode is supported in
LTE. In broadcast mode, a data stream is transmitted from a single source to multiple UEs in
the associated broadcast service area. In multicast mode (only relevant in case of 2G/3G), a
data stream is transmitted from a single source to UE(s) that belong to a multicast group in the
service area. In multicast mode, only users that are subscribed to the specific multicast service
and have joined the multicast group associated with this service can receive data. In broadcast
mode, users are not required to join or activate the service in order to receive the data.

MBMS was first specified for GPRS/UMTS in Release 6. To support flat architectures and
bypass the SGSN, Rel-8 introduced IP multicast as an option for the distribution of MBMS
payloads within the backbone network between GGSN and RNC. Each RNC wishing to
receive MBMS data needs to join a corresponding multicast group. The support for MBMS
in LTE/SAE was not included in Release 8 because of low interest in the industry.

Release 9 showed increasing interest on MBMS for LTE/SAE and specification work started
(see 3GPP TS 23.246 [47]). However, the target design for Release 9 EPS functionality was
limited to enable Mobile TV and scheduled file downloads. Therefore, MBMS for EPS (called
Evolved MBMS or shortly eMBMS) supports only MBMS broadcast mode.

Figure 1.18 shows the evolution of MBMS from Release 6 GPRS/UMTS to Release 8
UTRAN and Release 9 E-UTRAN.

The Release 9 MBMS broadcast mode function in E-UTRAN differs from GERAN/UTRAN
in that E-UTRAN does not support counting of active users in a cell. As a consequence, data
are broadcasted to predefined areas regardless whether there are any UEs in this area.

In E-UTRAN IP multicast is the only way for eNodeB(s) to receive MBMS data streams
(see Figure 1.18). In UTRAN the RNC may accept or reject IP multicast distribution and the
SGSN can establish normal MBMS point-to-point connections to all related RNC(s).

Evolved MBMS refers to the MBMS feature for EPS (specification started in 3GPP Release
9). New functional elements were introduced with eMBMS: The MBMS Gateway (MBMS

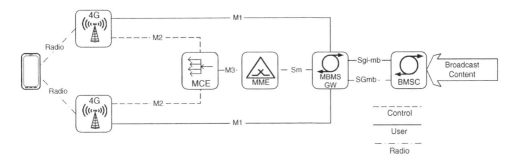

Figure 1.19   Evolved MBMS architecture

GW) replacing the 2G/3G GGSN and the Multicell Coordination Entity (MCE) used in E-UTRAN for uniform MBMS radio resource allocation and control of a group of cells.

Figure 1.19 depicts the eMBMS architecture including these new functional entities.

1.12.2   Description of Functional Entities

1.12.2.1   Broadcast Multicast Service Center (BM-SC)

The Broadcast Multicast Service Center (BM-SC) includes functions for MBMS user service provisioning and delivery. It is the entry point for the content provider, used to authorize and initiate MBMS bearer services within the PLMN via SGmb interface and to schedule and deliver data transmissions via SGi-mb. The BM-SC authenticates, authorizes, and charges access requests from the content provider. The interface between BM-SC and content provider is not specified by 3GPP (except for Public Safety group calls, see chapter 5).

1.12.2.2   MBMS Gateway (MBMS GW)

The MBMS Gateway delivers packets to eNodeBs in configured MBMS service areas via M1 interface and provides MBMS session control signaling (session start/stop/update) toward E-UTRAN (to the MCE) via the MME on Sm interface. The MBMS GW consists of a control and user plane part (MBMS CP and MBMS UP). The MBMS GW may be stand-alone or colocated with other network elements such as the BM-SC, S-GW, or P-GW. The Sn interface between MBMS GW and S4-SGSN is not shown in Figure 1.19. It provides control plane signaling similar to Sm and is used to forward MBMS data in point-to-point mode using GTP.

1.12.2.3   Multicell/Multicast Coordination Entity (MCE)

The Multicell/multicast Coordination Entity (MCE) provides functions for MBMS admission control and MBMS radio resource allocation. It interfaces with the eNodeBs in E-UTRAN via M2 interface. The MCE can be a stand-alone entity or colocated with an eNodeB, in which case only cells controlled by that particular eNodeB can form a MBSFN area (see next sections). In principle the MCE can also be colocated with the MBMS GW.

Table 1.7 Reference points and protocols for MBMS

Reference point	Protocols	Specifications
M1	GTPv1-U	TS 29.281 [22], TS 36.445 [48]
M2	M2AP	TS 36.443 [49]
M3	M3AP	TS 36.444 [50]
Mz (only for GPRS/UMTS)	DIAMETER	TS 29.061 [10]
Sm	GTPv2-C	TS 29.274 [19]
SGmb	DIAMETER	TS 29.061 [10]
SGi-mb	IP unicast or multicast	TS 29.061 [10]

1.12.2.4   MME supporting MBMS

The MME is enhanced for MBMS to support MBMS session control signaling via M3 interface to the MCE. Not shown in the Figure 1.19 is the Mz interface between a BM-SC in HPLMN and a BM-SC in VPLMN (roaming case). Mz is currently supported for GPRS and UMTS only, but not for LTE/SAE.

1.12.2.5   UE supporting MBMS

The MBMS-capable UE needs to support additional functions for the activation and deactivation of the MBMS bearer service and special MBMS security functions (e.g., support of key distribution via the MICKEY protocol).

1.12.2.6   Summary of Reference Points and Protocols

Table 1.7 summarizes the reference points used for MBMS.

1.12.3   MBMS Enhancements

In Release 10, MBMS was enhanced so that the network is capable to manage individual MBMS services depending on the number of users interested in a service. This enables prioritization of different MBMS services depending on their relative priority when there is resource shortage. In addition, a MBMS counting function was introduced to allow counting UE(s) in connected mode, either receiving a particular MBMS service or just interested in receiving a service. Note that only Release 10 devices in connected mode are counted. Release 10 devices in idle mode and Release 9 or older devices are not counted. The MBMS counting function is controlled by the MCE and allows the MCE to enable or disable MBSFN transmission for the service. In support of these new MCE functions, new Release 10 M2 interface procedures were introduced. These procedures support, for example, suspend and resume of a MBMS service, send a MBMS counting request, and obtain MBMS counting results. The prioritization of different MBMS services is also done by the MCE because it is responsible for controlling the allocation of radio resources for MBSFN transmission. So the MCE can pre-empt radio resources used by an ongoing MBMS service according to the Allocation and Retention Priority (ARP) of different MBMS radio bearers.

In LTE Release 11, MBMS was again enhanced to ensure MBMS service continuity in a multicarrier network deployment. MBMS services may be deployed on different carrier frequencies over different geographic areas. Release 11 enhancements allow the network to signal assistance information to MBMS-capable devices that provide information related to the actual MBMS deployment such as carrier frequencies and service area identities. In Release 11, a MBMS-capable device can indicate its interest in MBMS services by indicating the carrier frequencies associated with the MBMS services of interest and the priority between MBMS and unicast service. The network uses this indication for mobility management decisions so that the device is always able to use its receiver at the appropriate carrier frequency layer, thus ensuring continuity of MBMS services. In idle mode, a MBMS-capable device can prioritize a particular carrier frequency during cell reselection depending on the availability of MBMS services on that carrier frequency. To ensure MBMS service continuity in connected mode, the MBMS interest indication received from the device is signaled to the target cell as part of the handover preparation procedure.

1.12.4 MBSFN and MBMS Radio Channels

For the MBMS broadcast mode, E-UTRAN supports the so-called Multimedia Broadcast Single Frequency Network (MBSFN) feature where cells of an MBSFN area are synchronized and produce identical transmissions. MBSFN areas can be predefined. The MCE is in charge of uniform radio resource allocation and synchronized data delivery. The resulting signal will appear to a UE as just one transmission over a time-dispersive radio channel. Multiple cells can belong to a MBSFN area and every cell can be part of up to eight areas. Up to 256 different areas can be defined. It is also possible that certain cells in or at the edge of a MBSFN area do not support MBMS transmission, thus do not belong to the MBSFN area, but transmit other data with low power not to interfere with the MBMS signal. Such cells are referred to as "reserved cells." Figure 1.20 shows a MBMS Single Frequency Network configuration.

MBMS in E-UTRAN requires also new logical, transport, and physical channels. Two logical channels are related to MBMS.

The Multicast Traffic Channel (MTCH) carries data of a certain MBMS service. MBMS services in a MBSFN area may use multiple MTCH. As there is no feedback in the uplink, MTCH uses unacknowledged mode for data transmission. The Multicast Control Channel (MCCH) provides control information to receive MBMS services. There is one MCCH per MBSFN area. One or several MTCH and one MCCH are multiplexed at the MAC layer onto the Multicast Channel (MCH), which is multiplexed to the Physical Multicast Channel (PMCH). The MCCH provides information like the subframe allocation and modulation/coding scheme of each MCH. MCCH can also be used in unacknowledged mode. A notification mechanism is used to announce MCCH changes to the UE. Changes to the MCCH that are not announced can be detected by monitoring the MCCH at each modification period.

The transport format is determined by the MCE and signaled via the MCCH to the UE. During one MCH Scheduling Period (MSP), the different MTCHs and optionally the one MCCH are multiplexed on the MCH. The MCH Scheduling Information (MSI) is provided by the eNodeB at the beginning of the MSP to indicate which subframes are used by each MTCH during the MSP.

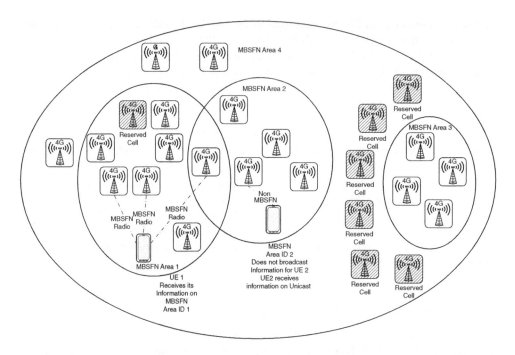

Figure 1.20   MBMS single frequency network

System information broadcast messages carry information on common and shared channels in E-UTRAN and provide also information related to MBMS transmission. They indicate which radio frames contain subframes that can be used for MBMS.

The role of the Broadcast Control Channel (BCCH) is to indicate MCCH-related resource information to the UE (for each MCCH in the cell independently). This information could be the scheduling of the MCCH for multicell transmission on the MCH, the MCCH modification period, the repetition period radio frame offset, and the subframe allocation.

1.13   Terms and Definitions

1.13.1   Roaming

By roaming a subscriber having a contract with network operator A can use network resources of another operator B. Operator A's network is called the Home PLMN (HPLMN) of the subscriber while operator B's network is called the Visited PLMN (VPLMN). Usually operator B's network is deployed in a country different to operator A's network. However, it is also possible that both networks operate in the same country. In the latter case we speak of "national roaming," otherwise we speak of "international roaming." National roaming allows an operator to increase radio coverage by using another operator's network while the latter operator benefits of additional roaming fees. As an example, Public Safety networks based on the LTE standard can increase their nationwide coverage by signing national roaming agreements with commercial LTE network operators in this country (another way to accomplish this is by sharing radio base stations as described in Section 1.11).

Roaming requires proper agreements between operators (so-called roaming agreements). These agreements clarify how operator A's subscriber can use operator B's network, which services the subscriber can use, and what roaming fees have to be paid. As a prerequisite for roaming, a UE that roams into a foreign network (a roaming-in UE) must support the radio technology and frequencies of this network and the user must have a subscription for the provided radio technology. As an example: If a UE is LTE capable and roams into a foreign LTE network, it cannot register with this network as long as the user has no valid LTE subscription data in his home network. Registration in a foreign network requires that the UE is provisioned with a list of possible/preferred roaming networks. This provisioning process is done by the home operator. On the basis of the list of roaming partners the UE can select the preferred network in a certain country. During the registration process, the HPLMN has to provide authentication and subscription data to the VPLMN via a signaling connection. Although data traffic can be routed by the VPLMN directly to the desired destination, it is common practice that user data are routed from the VPLMN to the HPLMN before traveling toward their final destination (e.g., to a Web server in the Internet). This allows the home operator to apply individual charging rules to the subscriber. For use of visited network resources the subscriber has to pay additional roaming fees on top of the usual service fees. Thus, the VPLMN has to report resource usage of HPLMN's subscribers to the home network (e.g., on a monthly basis).

1.13.2   Circuit-Switched and Packet-Switched Networks

In CS networks a connection (e.g., a voice call) can use a dedicated transmission channel with a constant data rate. This transmission channel can only be used by this particular connection, irrespective of whether data are transmitted or not. Examples for CS networks are the PSTN and networks based on the GSM standard.

In PS networks data that have to be transmitted are separated into data packets and each data packet is transmitted independently from the source to the destination. In principle different data packets can be routed via different paths. Transmission can be done in a connection oriented or connection less manner. Examples for PS networks are IP networks, for example, the Internet. While networks based on the UMTS standard have both a CS and PS component (called CS and PS domains), the new LTE standard consists only of a PS domain.

1.13.3   Access Stratum and Non-Access Stratum

The Latin word "Stratum" means layer and was chosen to avoid confusion with other layers such as the Open Systems Interconnection (OSI) layers.

The Access Stratum (AS) layer consists of all functions that are directly related to the radio access network and the control of connections between end user device and radio network. Protocols on AS layer run between the device and the radio base station in order to establish and maintain radio channels.

The NAS layer, on the other hand, is on top of the AS layer and consists of functions that are related to call control, mobility, and session management. Protocols on the NAS layer are exchanged between the device and the core network, that is, the NAS layer is transparent to

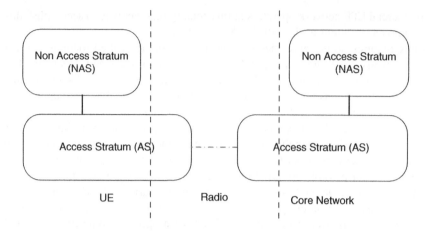

Figure 1.21   AS and NAS layer

the radio access network. The NAS layer and AS layer in the device and core network are able to communicate with each other. This allows the NAS layer to trigger establishment or termination of radio bearers that are used to exchange signaling or user data.

Figure 1.21 provides a simplified overview of the two layers.

References

[1]   3GPP TR 23.882: "3GPP System Architecture Evolution: Report on Technical Options and Conclusions".
[2]   3GPP TS 23.401: "GPRS Enhancements for E-UTRAN Access".
[3]   3GPP TS 23.402: "Architecture Enhancements for Non-3GPP Accesses".
[4]   3GPP TS 23.060: "General Packet Radio Service (GPRS); Service Description".
[5]   IETF RFC 5213: "Proxy Mobile IPv6".
[6]   IETF RFC 5555: "Mobile IPv6 Support for Dual Stack Hosts and Routers".
[7]   3GPP TS 36.104: "Base Station (BS) Radio Transmission and Reception".
[8]   3GPP TS 29.060: "General Packet Radio Service (GPRS); GPRS Tunnelling Protocol (GTP) across the Gn and Gp Interface".
[9]   3GPP TS 21.905: "Technical Specification Group Services and System Aspects; Vocabulary for 3GPP Specifications".
[10] 3GPP TS 29.061: "Interworking between the Public Land Mobile Network (PLMN) Supporting Packet-Based Services and Packet Data Networks (PDN)".
[11] 3GPP TS 23.203: "Policy and Charging Control Architecture".
[12] 3GPP TS 29.214: "Policy and Charging Control over Rx Reference Point".
[13] 3GPP TS 24.008: "Mobile Radio Interface Layer 3 Specification; Core Network Protocols".
[14] 3GPP TS 24.301: "Non-Access-Stratum (NAS) protocol for Evolved Packet System (EPS)".
[15] 3GPP TS 23.122: "Non-Access-Stratum (NAS) Functions Related to Mobile Station (MS) in Idle Mode".
[16] 3GPP TS 36.300: "Evolved Universal Terrestrial Radio Access (E-UTRA) and Evolved Universal Terrestrial Radio Access (E-UTRAN); Overall Description".
[17] 3GPP TS 36.413: "S1 Application Protocol (S1AP)".
[18] 3GPP TS 36.423: "X2 application protocol (X2AP)".
[19] 3GPP TS 29.274: "Evolved General Packet Radio Service (GPRS) Tunnelling Protocol for Control Plane (GTPv2-C)".
[20] IETF RFC 3588: "Diameter Base Protocol".
[21] 3GPP TS 29.272: "Evolved Packet System (EPS); Mobility Management Entity (MME) and Serving GPRS Support Node (SGSN) Related Interfaces Based on Diameter Protocol".

[22] 3GPP TS 29.281: "General Packet Radio System (GPRS) Tunnelling Protocol User Plane (GTPv1-U)".
[23] 3GPP TS 29.275: "Proxy Mobile IPv6 (PMIPv6) Based Mobility and Tunnelling protocols; Stage 3".
[24] 3GPP TS 29.215: "Policy and Charging Control (PCC) Over S9 Reference Point; Stage 3".
[25] 3GPP TS 29.212: "Policy and Charging Control (PCC); Reference Points".
[26] 3GPP TS 33.401: "3GPP System Architecture Evolution (SAE); Security Architecture".
[27] 3GPP TS 33.402: "3GPP System Architecture Evolution (SAE); Security Aspects of Non-3GPP Accesses".
[28] 3GPP TS 32.240: "Charging Architecture and Principles".
[29] 3GPP TS 32.251: "Packet Switched (PS) Domain Charging".
[30] 3GPP TS 32.296: "Online Charging System (OCS) Applications and Interfaces".
[31] 3GPP TS 32.273: "Multimedia Broadcast and Multicast Service (MBMS) Charging".
[32] IETF RFC 3261: "SIP: Session Initiation Protocol".
[33] IETF RFC 4566: "Session Description Protocol".
[34] 3GPP TS 24.229: "IP Multimedia Call Control Protocol Based on Session Initiation Protocol (SIP) and Session Description Protocol (SDP); Stage 3".
[35] 3GPP TS 29.229: "Cx and Dx Interfaces Based on the Diameter Protocol; Protocol Details".
[36] 3GPP TS 29.163: "Interworking between the IP Multimedia (IM) Core Network (CN) Subsystem and Circuit Switched (CS) Networks".
[37] 3GPP TS 23.216: "Single Radio Voice Call Continuity (SRVCC)".
[38] 3GPP TS 23.272: "Circuit Switched (CS) fallback in Evolved Packet System (EPS)".
[39] 3GPP TS 23.204: "Support of Short Message Service (SMS) over Generic 3GPP Internet Protocol (IP) Access".
[40] 3GPP TS 29.273: "Evolved Packet System (EPS); 3GPP EPS AAA Interfaces".
[41] 3GPP TS 24.302: "Access to the 3GPP Evolved Packet Core (EPC) via Non-3GPP Access Networks; Stage 3".
[42] 3GPP TS 24.244: "IP Multimedia Call Control Protocol Based on Session Initiation Protocol (SIP) and Session Description Protocol (SDP); Stage 3".
[43] IETF RFC 3748: "Extensible Authentication Protocol (EAP)".
[44] IETF RFC 5448: "Improved Extensible Authentication Protocol Method for 3rd Generation Authentication and Key Agreement (EAP-AKA′)".
[45] 3GPP TS 24.303: "Mobility Management Based on Dual-Stack Mobile IPv6; Stage 3".
[46] 3GPP TS 23.251: "Network Sharing; Architecture and Functional Description".
[47] 3GPP TS 23.246: "Multimedia Broadcast/Multicast Service (MBMS); Architecture and Functional Description".
[48] 3GPP TS 36.445: "Evolved Universal Terrestrial Radio Access Network (E-UTRAN); M1 data transport".
[49] 3GPP TS 36.443: "Evolved Universal Terrestrial Radio Access Network (E-UTRAN); M2 Application Protocol (M2AP)".
[50] 3GPP TS 36.444: "Evolved Universal Terrestrial Radio Access Network (E-UTRAN); M3 Application Protocol (M3AP)".

2

Regulatory Features

2.1 Emergency Calls

2.1.1 Overview

The support for emergency calls, that is, calls to a special emergency center, is a mandatory functionality in all cellular wireless networks. Strictly speaking, User Equipment (UE) support of emergency calls is only required under the condition that the UE supports also speech calls. So a data card UE with no speech call support does not need to support emergency calls. In principle, emergency services can use other means for communication as well, for example, text messages for disabled persons or video calls.

Emergency calls will be routed to a call center called PSAP (Public Safety Answering Point). In general, a call is routed to a particular PSAP based on the dialed number (such as 911 or 112) and location of the user in accordance with national regulations. There are different emergency call centers for police, ambulance, fire brigade, marine guard, and mountain rescue depending on the type of emergency. Emergency call centers may be connected to the Public Switched Telephone Network (PSTN), Circuit-Switched (CS) domain, Packet Switched (PS) domain, or any other packet network. For historical reasons they are still connected to the PSTN but a transition to IP-based PSAPs is underway in some countries. An emergency call over the Long-Term Evolution (LTE) Internet Protocol (IP)-based network utilizes IP Multimedia Subsystem (IMS) as application over the top. Both LTE and IMS required some adaptations to fulfill regulatory requirements and enable emergency calls.

2.1.2 Requirements

Subject to local regulations, emergency calls are prioritized over nonemergency calls by the network. National regulations may also require wireless networks to provide emergency caller's location to the dispatcher. Emergency calls may be supplemented with emergency-related data. Support of so-called PSAP callback sessions, that is, the network recognizes that a call is a callback from the PSAP to the originator of the emergency call (e.g., to get more information about an accident) and handles this call differently compared to normal calls, is also subject to local regulation.

LTE for Public Safety, First Edition. Rainer Liebhart, Devaki Chandramouli, Curt Wong and Jürgen Merkel.
© 2015 John Wiley & Sons, Ltd. Published 2015 by John Wiley & Sons, Ltd.

National regulations may decide whether networks should accept emergency calls from unauthenticated UEs or UEs without an Universal Subscriber Identity Module (USIM), thus networks must provide the ability to either allow or not allow emergency calls without USIM. As USIM-less emergency calls were misused by people in the past to test a mobile phone or dial fake emergency calls, these kinds of calls are no longer allowed in most countries. To support emergency calls without USIM or emergency calls placed by UEs that are connecting to a network they are not allowed, a special emergency registration and attach procedure was added to LTE. This allows UEs to place emergency calls even when normal registration would be rejected by the network. UEs can be divided in the following categories depending on which requirements must be fulfilled to place an emergency call over LTE:

a. **Normal service UE(s).** First category of UE(s) are the ones that have a valid subscription, can be authenticated and authorized for PS services and in addition, they are able to perform IMS registration in the registered location. If not attached already, such UEs perform a normal attach as they would do for any other call and initiate a UE requested Packet Data Network (PDN) connectivity procedure for emergency services with request type "emergency" when they detect that the user tries to make an emergency call. Normal attach will allow the UE to obtain regular services in addition to emergency services unlike emergency attach which allows the UE to obtain emergency services only.
b. **Only UEs that can be authenticated are allowed.** These UEs must have a valid USIM. These UEs are authenticated but may be in a so-called limited service state due to being in a location where they are not allowed to obtain service. UEs that cannot be authenticated will be rejected.
c. **UEs that can be identified but authentication is optional.** These UEs must have an USIM. If authentication fails for whatever reason, the UE is in limited service state but is granted access and the unauthenticated International Mobile Subscriber Identity (IMSI) is retained in the network for recording purposes. The network must request the device identity International Mobile Equipment Identity (IMEI) and use it as device identifier. UEs without USIM or without an unlocked USIM will be rejected.
d. **All UEs are allowed.** Along with authenticated UEs, this includes UEs with an IMSI that cannot be authenticated (e.g., due to an expired subscription or no valid roaming agreement) and UEs with only an IMEI (e.g., no USIM or USIM cannot be unlocked by the entered PIN code). This is again referred to as UEs in limited service state. If an unauthenticated IMSI is provided by the UE, the unauthenticated IMSI is retained in the network for recording purposes. The IMEI is used in the network to identify the device.

2.1.3   Emergency Call Architecture

Emergency calls in LTE are supported using either IMS or CS fallback. The IMS architecture for supporting emergency services is defined in the 3rd Generation Partnership Program (3GPP) Technical Specification (TS) 23.167 [1] and the Evolved Packet System (EPS) level aspects in 3GPP TS 23.401 [2]. The CS fallback architecture is specified in 3GPP TS 23.272 [3]. In current Global System for Mobile Communications (GSM)/Wideband Code Division Multiple Access (WCDMA)/Code Division Multiple Access (CDMA) networks, emergency calls are supported via the CS domain. CS Fallback procedure was defined for LTE in 3GPP since Release 8. Emergency calls can be supported in LTE with CS fallback toward GSM/WCDMA (CDMA case) networks. If Circuit-Switched Fall Back (CSFB) is not supported, the UE can autonomously switch to a neighboring GSM/WCDMA/CDMA network to initiate emergency calls. IMS Voice over IP (VoIP) emergency call support over LTE has been specified in 3GPP Release 9.

2.1.3.1 IMS Emergency Calls

Emergency services are provided locally by the serving network, that is, whenever a user roams into a foreign country and dials a well-known emergency number the call is routed to the nearest PSAP in this country. As voice services over LTE may utilize IMS (known as "VoLTE" service), IMS and LTE must support emergency calls to fulfill regulatory requirements. This section provides more insights into the architecture for IMS emergency calls over LTE and related procedures. High-level architecture is shown in Figure 2.1.

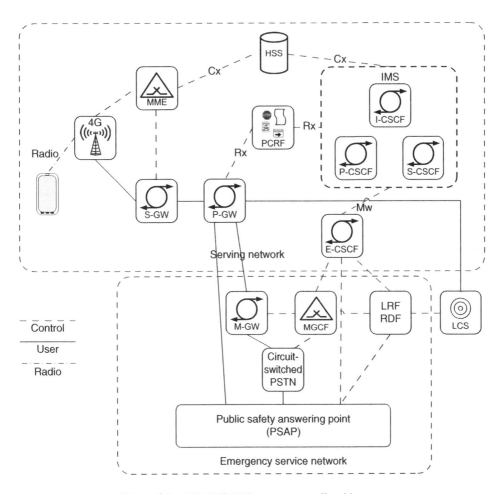

Figure 2.1   LTE IMS VoIP emergency call architecture

In contrary to normal voice calls over IMS, emergency calls over IMS require special treatment in the device, network (LTE/System Architecture Evolution (SAE)), and on the application layer (IMS). We have hinted some of these special requirements in the previous sections but here is a summary again:

- Emergency calls are routed locally in the (visited or roaming) network to the next available PSAP or emergency service center.
- In some countries, a PSAP must be able to callback the originator of the emergency call.
- The network must be able to provide accurate information about caller's location to the PSAP.
- Depending on the local regulation, devices in limited service state that are not allowed to connect to the whole network or not in a certain area for normal services must be able to make an emergency call.
- Depending on the local regulation, devices without USIM must be able to make an emergency call.
- Emergency calls deserve prioritized handling, for example, when the network is in a congestion situation or in overload.
- In order to allow a clear cost split between emergency and other services, resources reserved for an emergency connection are not allowed to be used for other services such as normal voice calls. By doing so, the operator can decide how to charge emergency calls. For example, depending on the country, the service may be sponsored by the national government or the wireless service provider. In some cases, wireless service providers may choose to pass on their costs of providing emergency services to their end customers.

The listed general requirements translate into the following high-level roles and responsibilities of devices and network elements during an emergency call setup.

UE

A UE can either camp normally in a suitable cell (i.e., selected cell is where normal services are offered) or camp in "limited service" state in the selected cell. If the UE is unable to find a suitable cell to camp on (i.e., selected cell is unable to provide normal service due to reasons such as lack of subscription, roaming agreement, etc.), then it enters "limited service" state. Detailed cell selection criteria and idle mode camping procedures for the UE are specified in 3GPP TS 36.304 [4] and 3GPP TS 23.122 [5].

UEs camping normally in a suitable cell needs to perform the following two main functions:

1. Initiate normal attach, if not attached already.
2. Initiate a special UE-requested PDN connectivity procedure for emergency services. The PDN connection request carries a special request type set to "emergency," which allows the network to detect request for an emergency connection quite easily. In addition, the UE shall not include an APN (Access Point Name) when the request is for emergency.

UE(s) camping in "limited service" state in the selected cell needs to perform the following two main functions:

1. Initiate a special emergency attach procedure. The network can prioritize an emergency attach in case of congestion and suppress subscription checking and authentication/authorization procedures.

2. Initiate a special UE-requested PDN connectivity procedure for emergency services with emergency request type but no APN (this step is the same as for UEs camping normally).

eNodeB

It broadcasts support for IMS emergency calls to all UEs in coverage. It prioritizes the establishment of radio resources for requests based on emergency indicator. Prioritizes user plane establishment and retains existing user plane for emergency services based on QoS Class Identifier (QCI) and Address Resolution Protocol (ARP) in case of resource contention.

MME

Mobility Management Entity (MME) prioritizes the Attach and PDN connectivity requests from UEs when request is marked for emergency usage. MME stores emergency configuration data for use when UE requests for emergency bearer services. It retains emergency bearer services in case of overload and ignores certain restrictions for emergency-attached UEs (e.g., based on UEs location).

PDN-GW

When needed, Packet Data Network Gateway (PDN-GW) should have the ability to support initiating dedicated bearer establishment when an IMS VoIP emergency call is being established. PDN-GW starts a configurable inactivity timer (e.g., to enable PSAP call back) for the PDN connection when informed by the Policy and Charging Rules Function (PCRF) about an emergency service session.

PCRF

When informed by the Proxy Call Session Control Function (P-CSCF) that a new session is for emergency usage, PCRF removes all PCC rules with a QCI other than the default bearer QCI and the QCI used for IMS signaling. It also updates the PDN-GW.

MGCF/MGW

When converting a Session Initiation Protocol (SIP) to an ISDN User Part (ISUP) message within an emergency call, Media Gateway Control Function (MGCF) may indicate that the ISUP message (e.g., initial address message to setup the call) is high priority or for an emergency call. Such indications are normally country specific.

P-CSCF

Proxy CSCF has a list of emergency numbers in a certain country or worldwide known countries configured. Once an IMS emergency call is detected based on the received SIP URI or Tel URI, the P-CSCF forwards the call to the E-CSCF (see below) and informs the PCRF about a new session used for emergency services.

S-CSCF

Serving S-CSCF acts as the registrar for the IMS emergency registration including authentication. The S-CSCF is not in the call chain of the emergency call itself.

E-CSCF

Emergency Call State Control Function (E-CSCF) is a specific CSCF that takes the call handling role from the S-CSCF for emergency calls since only they require specific processing rules, for example, retrieving location information or disabling user-specific services such as call forwarding. E-CSCF determines the correct PSAP based on location information and routes emergency calls toward PSAP.

LRF

Location Retrieval Function (LRF) retrieves the location of the UE from the Location Server (LCS) and stores it. E-CSCF can interrogate the LRF for routing of emergency calls to the closest PSAP and for dispatching the emergency services to the right address.

RDF

Routing Determination Function (RDF) is usually integrated in a LCS or LRF, provides the proper PSAP destination address to the E-CSCF for routing the emergency request, that is, the RDF provides directly a routable SIP URI of the PSAP to E-CSCF. If there is no RDF in place, E-CSCF needs to translate location information and emergency number into a routable SIP address on its own.

Location Server (LCS)

Obtains the location of the UE. The LCS interacts with other network elements such as MME and/or eNodeB to retrieve UE's location. Location information can be in the simplest case just the cell ID from which the UE connects to the network or it can also be more accurate $X-Y$ coordinates.

PSAP

A PSAP is a kind of call center responsible for answering calls destined to an emergency telephone number.

Procedure for IMS Emergency Calls

The signaling procedure for normally attached UEs for establishing an emergency call is not much different to a normal call setup procedure. The emergency call setup procedure for UEs in "normal service" state is shown in Figure 2.2.

The emergency call setup procedure for UEs in "limited service" state, shown in Figure 2.3, is by nature different from the procedure for UEs in "normal service" state.

An UE that can normally attach to the network initiates a PDN connectivity request procedure in order to perform an emergency call (reference Figure 2.2). An UE that intends to perform an emergency call but cannot attach normally due to either roaming restrictions or lack of valid credentials or lack of USIM performs an emergency attach by setting the attach type as emergency in the request (refer Figure 2.3).

The UE must identify itself during the attach procedure, both normal and emergency attach. The mobile identity can be a temporary identity assigned by the network or the subscription-based IMSI read from the USIM. For security reasons IMSI is used only, if no temporary identity exists.

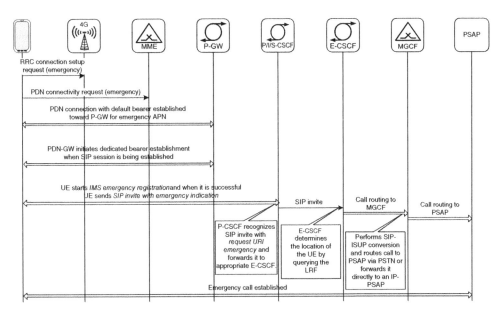

Figure 2.2 LTE IMS VoIP emergency call for normal UEs

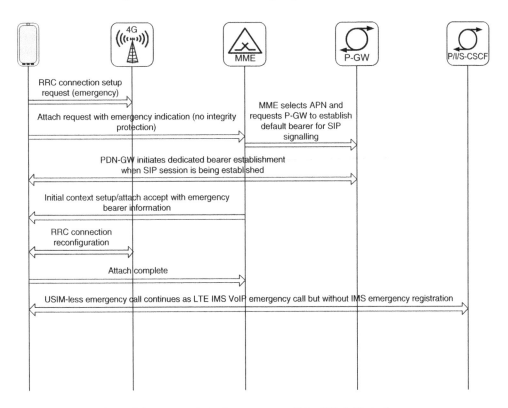

Figure 2.3 LTE IMS VoIP emergency call for UEs in limited state

When a device has got no USIM and therefore has none of the above-mentioned identities, then it can use its device identity (IMEI) in the emergency attach as the last resort. In this case, if the MME supports emergency attach with IMEI (e.g., based on local regulation), the MME skips authentication, ciphering, and integrity protection of those UEs that do not have security credentials and accepts attach.

If the MME does not support USIM-less emergency calls (e.g., local regulation does not allow this), receiving the IMEI as mobile identity triggers the MME to reject the emergency attach with IMEI and indicate to the UE that IMEI-based attach is not accepted.

The network stores emergency configuration data that consists of: QoS profile, Access Point Name-Aggregate Maximum Bit Rate (APN-AMBR), emergency APN, and statically configured P-GW. User subscription data will be ignored for emergency bearers. Unauthorized users may not have any subscription data.

For emergency use, the network can accept EPS Attach or Tracking Area Update (TAU) coming from an area that is forbidden for the UE, although normal services for the UE are still not allowed in that area.

If the UE has successfully registered to the network and has obtained emergency bearer resources when moving to an area where the UE is not allowed normal services by subscription, then after TAU the network must tear down all normal bearer resources, but maintain the emergency bearer resources. This case is related with regional roaming restrictions and the UE roaming outside of its regionally provisioned area in an allowed Public Land Mobile Network (PLMN). The logic behind is that emergency call does not allow the user to stretch other services to areas that are not allowed otherwise.

Bearer resource modification is not allowed for any PDN connection that has been established for emergency services. If the network receives such request from the UE, it must reject it. Default and dedicated bearers of a PDN connection associated with the emergency APN are always dedicated for emergency sessions. Additional PDN connections (i.e., besides the emergency PDN connection) are not allowed for UEs that are emergency attached. This is reasonable as the user does not have a proper subscription with the operator and the operator is not obligated to provide other services to the user except emergency services mandated by the national regulator.

Emergency calls are protected against network-initiated detach. If the UE has an emergency bearer context active, network-initiated detach requested by the Home Subscriber Server (HSS) should result in deactivation of all nonemergency bearer contexts. The network also attempts to maintain support for emergency bearer services in overload situations.

Upon successful establishment of emergency bearers, the UE may perform IMS emergency registration (as a prerequisite the UE needs to have all necessary credentials for IMS registration) (refer 3GPP TS 23.167 [1]). The UE performs IMS emergency registration in the following scenarios:

– If the UE is currently attached to its home operator's network (HPLMN, Home Public Land Mobile Network) and IMS registered, but the network indicates its support for emergency bearers.
– If the UE is currently attached to its home operator's network (HPLMN) but is not IMS registered.
– If the UE is attached to a different network, that is, not its home operator's network (Visited Public Land Mobile Network (VPLMN) in the home or foreign country). In this case, the UE may or may not be IMS registered with its HPLMN.

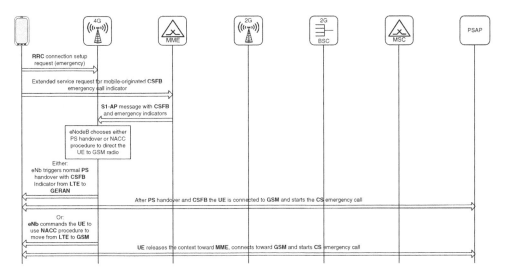

Figure 2.4   CSFB emergency call – PS Handover/NACC

The UE does not perform an IMS emergency registration in the following cases:

– If the UE has no credentials, the UE establishes an IMS emergency call without IMS registration (support for this depends on local regulation).
– If the UE is currently attached to its home operator's network (HPLMN) and IMS registered, but the network does not support emergency bearers (this scenario depends on home network operator policies in which case the emergency call is treated like a normal call from EPS perspective). In this case, UE sets up an emergency call using the existing regular IMS registration.

The P-CSCF is configured to recognize an emergency call session and forwards the call set up to an appropriate E-CSCF. If available, the E-CSCF queries the LRF for location information. Afterwards the E-CSCF forwards the call to the proper PSAP either directly or via a MGCF.

2.1.3.2   CSFB Emergency Calls

Since not all LTE networks may deploy IMS and support voice (VoIP) services, 3GPP agreed to define CS fallback toward GSM/EDGE Radio Access Network (GERAN) and Universal Terrestrial Radio Access Network (UTRAN) network to support voice. CS fallback can be supported either using PS handover or network-assisted cell change. CS fallback to UTRAN is supported only using PS Handover while CS fallback to GERAN is supported using both PS handover and Network-Assisted Cell Change (NACC). These methods are illustrated in Figure 2.4.

When an emergency call is initiated the UE connects to the network by indicating "emergency" (refer TS 36.331 [6]). After it is successfully connected, UE sends a request for CSFB emergency call to the network. The MME redirects the UE toward 2G/3G network by instructing the eNB. Depending on the target cell, eNB triggers either PS handover or NACC procedure. If the Evolved NodeB (eNB) decides to perform PS handover, it triggers a normal PS handover procedure with CSFB indicator from LTE to UTRAN or GERAN. If eNB decides to trigger NACC, it commands the UE to move from LTE to GERAN. Upon completion of PS handover or cell change, UE connects to UTRAN or GERAN. Then, the UE initiates a CS emergency call setup in the UTRAN/GERAN CS domain.

2.1.4   PSAP Callback

The capability of the person in the PSAP to call back a UE, for example, in case the connection was interrupted has not been fully addressed in 3GPP. So, a PSAP callback is treated just like a normal call by the network. However, service requirements for PSAP callback exist since 3GPP release 7 in 3GPP TS 22.101 [7]. Actual definition in stage 3 and implementation of these requirements have been pending for completion of related work in Internet Engineering Task Force (IETF). IETF draft in Ref. [8] introduces "psap-callback" as another value for the SIP Priority header. This value indicates that the concerned SIP request is originated by the PSAP to a UE which had originated a call to that PSAP.

If no special PSAP callback indication is received, the call is to be treated as a normal voice call. PSAP or Emergency Centers are allowed to establish a call to which there had been an emergency call previously and the UE had registered with valid credentials. Indication that a call is a PSAP call back has to be supported when required by local regulation. If a terminating call has a PSAP call back indication, the UE should be able to detect this.

2.1.5   Emergency Numbers

If a user dials a local emergency number and the number is known to the UE, then the UE can detect that this is an emergency call. It is mandatory for the UE to store 112 and 911 among others as emergency numbers. There is multitude of emergency numbers used in countries across the globe. Hence, it is possible that a roaming UE does not detect all dialed numbers as an emergency number. If the dialed number is not detected by the UE as an emergency number, then the IMS network (P-CSCF) must recognize the number as emergency number and route the call to the next PSAP. Such a call will however use a normal PDN connection for IMS voice services.

Operators are allowed to add emergency numbers list to the UEs. This list contains numbers that are required for local needs. By default, the device must consider the following numbers as emergency numbers according to 3GPP TS 22.101 [7]:

1. 112 and 911 are always considered emergency numbers.
2. Any additional emergency numbers that are specified as such on the USIM.
3. 000, 08, 110, 118, 119, and 999 are considered as emergency numbers by a device without USIM.
4. Any additional emergency number that have been downloaded by the serving PLMN (VPLMN) to the device.

Although some countries may not use 112 or 911 as emergency numbers, the UE must always consider those as emergency numbers hence they cannot be used for any other purpose anywhere. Even if the national emergency number is, say 999, both 112 and 911 will also work as emergency numbers. Those inbound roamers who are used to the well-known numbers 112 and 911 at home can use them also when roaming in a foreign country.

In the countries where additional emergency numbers are used, the operators can configure the additional numbers on their subscribers' USIM.

2.1.6 Non Voice Emergency Services

For a typical emergency service, voice media is used to convey information between the caller and PSAP. However, there is a desire to allow individuals with disabilities (e.g., hearing impaired or speech disability) or under certain restricted situations (e.g., hiding from a suspect) to initiate an emergency session without a voice media or in addition to using voice media (e.g., sending a chat message, video, files, and so on while talking over the phone).

Emergency session in IMS without a voice component follows the same architecture and procedures as with voice sessions except that the media type will indicate that it is not voice related. The other allowed media types of IMS emergency sessions are real-time video (simplex or full duplex), text-based instant messaging, file transfer, video clip sharing, picture sharing, audio clip sharing, and Global Text Telephony (TTY).

Same as for emergency voice sessions, the use of nonvoice media is not a subscription-based service. Hence, a UE that is capable to establish IMS emergency sessions and supports other media types should also be capable to initiate IMS emergency session with media other than voice.

IMS emergency sessions can be initiated with both voice and nonvoice media together. If handover to a non-IMS voice supporting network (e.g., GSM) is required, then the voice media is maintained as CS emergency call but other nonvoice media will be removed.

2.1.7 Automated Emergency Calls

The eCall service is a European initiative intended to bring rapid assistance to motorists involved in a collision anywhere in the European Union. The eCall initiative aims to deploy a device installed in a vehicle that will automatically dial an emergency number (e.g., 112) in the event of a serious road accident to the closest emergency center, and wirelessly send airbag and sensor information, as well as GPS coordinates to local emergency agencies. The eCall service builds on the E112 emergency call. The European Commission is aiming to have a fully functional eCall service to be in place throughout the EU by 2015. It could be foreseen that eCall speeds up emergency response times significantly in urban areas and in rural areas.

The current eCall service is based on the circuit-switched emergency call in GSM and Universal Mobile Telecommunications System (UMTS) networks. eCall was deliberately designed in a way that works with any GSM or UMTS mobile network that deploys a CS domain. Thus, data are transferred in a circuit-switched channel using a special in-band codec during the first few seconds of the emergency call. Only after that, voice communication between the PSAP and the passengers in the vehicle is established. Figure 2.5 shows eCall information flow.

The longevity of GSM networks in the EU over the lifetime of vehicles is uncertain and GSM spectrum is likely to be reallocated for UMTS and/or LTE. In LTE, emergency

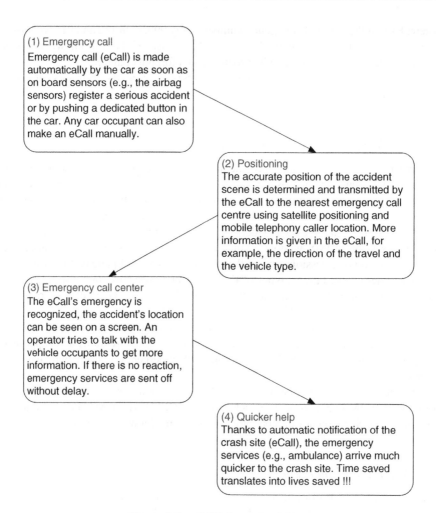

(1) Emergency call
Emergency call (eCall) is made automatically by the car as soon as on board sensors (e.g., the airbag sensors) register a serious accident or by pushing a dedicated button in the car. Any car occupant can also make an eCall manually.

(2) Positioning
The accurate position of the accident scene is determined and transmitted by the eCall to the nearest emergency call centre using satellite positioning and mobile telephony caller location. More information is given in the eCall, for example, the direction of the travel and the vehicle type.

(3) Emergency call center
The eCall's emergency is recognized, the accident's location can be seen on a screen. An operator tries to talk with the vehicle occupants to get more information. If there is no reaction, emergency services are sent off without delay.

(4) Quicker help
Thanks to automatic notification of the crash site (eCall), the emergency services (e.g., ambulance) arrive much quicker to the crash site. Time saved translates into lives saved !!!

Figure 2.5   eCall informational flow

calls can be placed either using CSFB or IMS. European Telecommunications Standards Institute (ETSI) Technical Committee Mobile Standards Group (TC MSG) has a work item on Migration of eCall Transport. As part of this study, Technical Report (see ETSI TR in Ref. [9]) was created. The scope of the document was to analyze the necessary adaptations of IMS emergency call and IMS Multimedia Emergency Service for supporting current and future services required by eCall, assessment of possible solution options, that is, in-band modem solution in case there is no use of CS bearers, Hybrid CS/IMS solution, and migration options. The IETF Emergency Context Resolution with Internet Technologies (ECRIT) is currently working on drafts for eCall [10] and Automatic Crash Notification (ACN) [11] contains information about activities outside Europe.

On the basis of input from ETSI [12], 3GPP has discussed about supporting eCall service based on VoLTE. When there is agreement to initiate such a work in 3GPP, the applicability of the existing technical solution for eCall (in-band modem) may be assessed as part of this work, as well as new technical solutions to be developed that are suitable for packet-switched networks such as UMTS and LTE.

2.2 Public Warning System

Public Warning System (PWS) is a service offered to local, regional, or national authorities to reach out to a public mass warning them of an impending emergency situation. Such warning messages may be necessary for a number of reasons [13] such as

- weather-related emergency such as tornados, ice storm, and hurricane;
- geological disasters such as earthquake and tsunami;
- industrial disasters such as the release of toxic gas or contamination;
- radiological disasters such as a nuclear plant disaster;
- medical emergencies such as an outbreak of a fast-moving infectious disease;
- warfare or acts of terrorism.

Some cities also use emergency warnings to advise people of prison escapes, abducted children, emergency telephone number outages, and other events.

LTE supports warning message delivery similar to the Cell Broadcast Service (CBS) in GERAN and UTRAN as specified in 3GPP TS 23.041 [14], but permits for a number of unacknowledged warning messages to be broadcasted to UEs within a particular service area. The term "unacknowledged" implies that there is no retry mechanism for warning messages.

3GPP Release 8 includes basic functionality to support the Earthquake and Tsunami Warning System (ETWS) feature that is based on Japanese requirements. It allows only one message broadcast at one time, a new message replaces the existing broadcast message with the newer one immediately. In Release 9 the functionality was enhanced to support the Commercial Mobile Alert System (CMAS) feature based on US requirements. Release 9 allows concurrent broadcast of multiple warning messages. In due course, new regional variants were introduced, new message identifiers have been defined, and the existing ones were enhanced. In Release 10 the Korean flavor called Korean Public Alert System (KPAS) was specified, and in Release 11 the European Public Warning System (EU-ALERT) was introduced. In Release 12, additional enhancements were introduced to improve robustness of the network (e.g., features to handle radio network failure, status reporting toward Cell Broadcast Center (CBC)); ability for the CBC to stop broadcasting of all messages in a certain area and in general, features to improve ease of network operation and maintenance. The following Figure 2.6 shows the architecture of PWS.

The CBC belongs to serving operator's domain. SBc is the reference point between CBC and MME for warning message delivery. The interface between CBC and CBE is out of scope of 3GPP. It is under the responsibility of public authorities and network operators in each country to agree on a national standard for this interface.

A simplified description of a warning message delivery is as follows:

1. The CBC identifies which MMEs need to be contacted and sends a request to MMEs. The request contains the warning message to be broadcasted, the warning area information, and the delivery attributes.

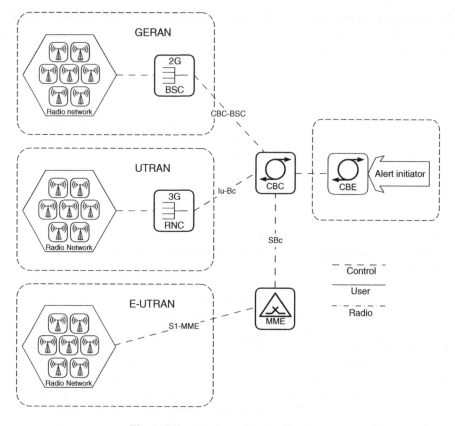

Figure 2.6   Warning system architecture

2. The MME forwards the warning message to eNodeBs and uses the Tracking Area ID list to determine the eNodeBs in the delivery area. If this list is empty, the message is forwarded to all eNodeBs that are connected to the MME.
3. The eNodeBs use the warning area information to determine the cell(s) where the message has to be broadcasted. If the cell ID information is not present, then the eNodeBs broadcast messages in all cells. The eNodeBs are responsible for scheduling the broadcast of messages and repetitions in each cell.

2.3   Lawful Interception

2.3.1   Principles

Lawful (or Legal) Interception (LI) in public telecommunication networks is a regulatory obligation in basically every country around the world. Therefore, interception is an integral capability of each modern telecommunication network. The process to order and execute

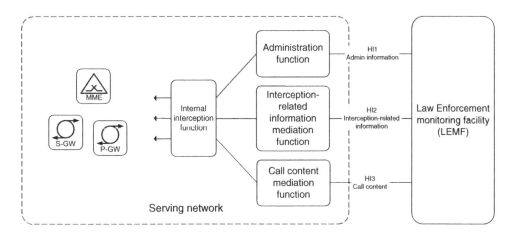

Figure 2.7   Lawful interception handover interfaces

interception is strictly regulated by national laws and telecommunication regulations to avoid misuse of this function.

Even though the national laws require different details of interception functionality, the principal architecture is the same for almost all countries and network domains.

The 3GPP LI architecture provides the Law Enforcement Agency (LEA), a government agency responsible for the enforcement of laws, with three interfaces by which Lawful Interception can be initiated and data being delivered to the LEA. These interfaces are called Handover Interfaces 1, 2, and 3 (HI1, HI2, and HI3) (shown in Figure 2.7). "Handover" in this context does not refer to handover of a UE from one cell to another due to mobility but just to provide content from the network domain to the LEA domain.

Handover Interface 1 (HI1) is used to deliver the interception warrant order to the telecommunication provider and to initiate LI for certain subscribers. The HI1 interface is typically a fax-/paper-based interface and not standardized. The warrant describes the interception target (e.g., Mobile Subscriber ISDN Number (MSISDN), IMSI, or IMEI), the length of the intercep-.tion period, and usually the delivery address of intercepted communication and events. These delivery addresses are used at the Handover Interface 2 (HI2) and Handover Interface 3 (HI3) along with other interception identifiers received with the warrant to send data to the LEA. The additional identifiers help the LEA to correlate interception data received via HI2 and HI3 interfaces regarding the same communication event.

HI2 delivers call-related data, for example, date, time, communicating party identifiers, call duration, used supplementary services, in a so-called Interception Record Information (IRI) event to the LEA. In case of a communication, an IRI contains correlation information enabling the LEA to identify the proper HI3 interception data.

HI3 can be used to provide a copy of the actual communication content (e.g., voice and video samples, data) to the LEA. This is valid for both circuit-switched and packed-switched communication. At the HI3 interface, correlation information is also delivered to the LEA to identify the matching HI2 IRI event reports. The LEA can instruct the network whether to deliver only call-related data or also the content.

Additional mediation functions in the network provide network hiding and postprocessing functionality for the IRIs and in some cases for the intercepted communication stream as well. They also duplicate the intercepted data toward several LEAs, if more than one LEA is intercepting the same target at the same time.

For security reasons, it is important to keep the interception warrant information, that is, who is going to be intercepted, secret. This information is not even accessible by network operator's maintenance personnel. Thus, eNodeB and UEs are not involved in the LI process as both of them are considered not trusted entities with respect to LI.

It has to be pointed out that LI mechanisms will also be applicable to Public Safety communication.

2.3.2   Lawful Interception for EPS

The Lawful Interception functionality for EPS was defined during 3GPP Release 8. The content of communication (in general IP packets) and interception-related information is provided by the EPS gateway nodes Serving Gateway (S-GW) or optionally Packet Data Network Gateway (P-GW). Additional header information is added to the intercepted packets. This extra header allows identifying the actual interception and correlates the HI3 interception data with the HI2 IRI event reports.

Additionally, the MME and the HSS also provide interception-related information to the LEA as they play an important role in mobility management procedures like handover and in case of roaming.

The EPS interception configuration architecture can be found in 3GPP TS 33.107 [15], while 3GPP TS 33.108 [16] specifies the encoding of intercepted content and interception-related information.

2.4   Enhanced Multimedia Priority Services

Enhanced Multimedia Priority Service (eMPS) (refer 3GPP TS 22.153 [17]) allows authorized service users to gain priority access to radio and core network resources when the network is congested. This creates the ability to deliver and complete service user's request with higher probability of success. The authorized service user has received a priority level assignment from an authorized agency (e.g., the local government), which has a prior arrangement with a mobile network operator that supports the eMPS service.

eMPS can be viewed as the evolution of Wireless Priority Service (WPS) because it defines the priority access mechanisms for packet-switched services and IMS services while WPS only deals with circuit-switched voice/data services.

While WPS only allows on demand type of invocation, eMPS also allows for "always-on" type of subscription as well. Figure 2.8 shows the overall eMPS architecture and its related parameters and interfaces.

The subscription data in HSS contain the IMS level priority parameter, which is specified by IETF RFC 4412 [18] and the related priority level. This information is passed to the P-CSCF by S-CSCF during IMS registration and allows the P-CSCF to set service user's IMS session

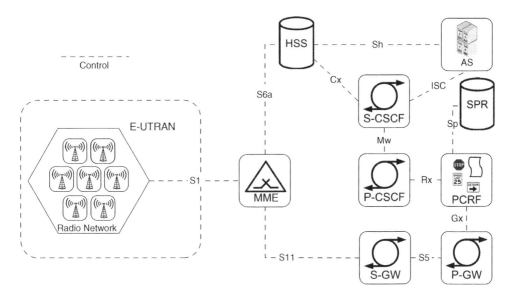

Figure 2.8   eMPS architecture

priority accordingly. A typical usage scenario is as follows: eMPS subscriber dials the destination number with a prefix (e.g., *272) in the front. On the basis of local configuration in the
Mobile Switching Center (MSC) Server or P-CSCF, serving network recognizes the prefix to
set special priority for eMPS subscriber.

An IMS Application Server can also retrieve this IMS priority setting via Sh. This is useful
for Application Server-based priority service detection (e.g., calling to certain 800 number).
The other subscription level parameter is MPS-CS, which indicates whether the UE can initiate "high-priority access" when invoking CSFB service while MPS-EPS indication is for the
native EPS service. The MME can use these indications from UE's subscription to determine
whether the priority request from the UE is legitimate.

eMPS-related subscription data can be categorized as follows:

- EPS Priority: If set to Yes it means the EPS bearer is given higher priority. If this parameter
  is always set to "Yes," then this corresponds to the "always on" setting, which means the
  EPS bearer is assigned a higher priority at the time of initial attach. If this parameter is
  allowed to be changed, then this is the "on demand" type of behavior, which means that
  the invocation is based on a request from the service user. For example, the service user
  invoking an application or a central agency remotely requests a priority upgrade.
- "IMS Signaling Priority": Means the IMS Signaling Bearer and the Default Bearer have
  higher priority at the time of IMS registration.
- "Priority Level": This parameter is used by the PCRF to determine the associated ARP value
  to be used at the EPS level.

The ARP parameter determines how the requested EPS resources (on radio and core level)
will be handled by the system. Each EPS bearer is associated with an ARP setting, which

indicates whether resources can be taken from lower priority bearers or be taken away by other higher priority requests once a resource shortage occurs in the network.

References

[1] 3GPP TS 23.167: "IP Multimedia Subsystem (IMS) Emergency Sessions".
[2] 3GPP TS 23.401: "GPRS Enhancements for E-UTRAN Access".
[3] 3GPP TS 23.272: "Circuit Switched (CS) Fallback in Evolved Packet System (EPS)".
[4] 3GPP TS 36.304: "Evolved Universal Terrestrial Radio Access (E-UTRA); Radio Resource Control (RRC); Protocol Specification".
[5] 3GPP TS 23.122: "Non-Access-Stratum (NAS) Functions Related to Mobile Station (MS) in Idle Mode".
[6] 3GPP TS 36.331: "Evolved Universal Terrestrial Radio Access (E-UTRA); User Equipment (UE) Procedures in Idle Mode".
[7] 3GPP TS 22.101: "Service Principles".
[8] IETF draft: "draft-ietf-ecrit-psap-callback".
[9] ETSI TR 103 140 "Mobile Standards Group (MSG), eCall for VoIP", http://www.etsi.org/deliver/etsi_tr/103100 _103199/103140/01.01.01_60/tr_103140v010101p.pdf.
[10] Next-Generation Pan-European eCall; draft-gellens-ecrit-ecall-01.txt
[11] Internet Protocol-based In-Vehicle Emergency Calls; draft-gellens-ecrit-car-crash-01.
[12] 3GPP Tdoc: C1-141868, "Migration of eCall Transport", ETSI MSG, Chairman Mr.Esa Barck, http://www.3gpp .org/ftp/tsg_CT/WG1_mm-cc-sm_ex-CN1/TSGC1_87_Phoenix/docs/C1-141868.zip.
[13] http://en.wikipedia.org/wiki/Public_warning_system
[14] 3GPP TS 23.041: "Technical Realization of Cell Broadcast Service (CBS)".
[15] 3GPP TS 33.107: "Lawful Interception Architecture and Functions".
[16] 3GPP TS 33.108: "Handover Interface for Lawful Interception (LI)".
[17] 3GPP TS 22.153: "Multimedia Priority Service".
[18] IETF RFC 4412: "Communications Resource Priority for the Session Initiation Protocol (SIP)".

3

LTE for Public Safety Networks

3.1 Why LTE for Public Safety Networks?

When the Federal Communications Commission (FCC) of the United States of America (USA) decided to use Long-Term Evolution (LTE) as the new radio technology for public safety communication, it was building on the assumption of a global widespread adoption of the LTE technology. At the time of writing this book, the Global Mobile Suppliers Association (GSA) had listed 300 deployed LTE networks in 107 countries and forecasts 350 networks by the end of 2014. Nearly 1900 LTE-capable devices are available on the market by now. For more information on market figures, see Ref. [1].

One important driver behind the desire to use LTE for Public Safety Communication was the economies of scale achievable with LTE compared to existing Public Safety Communication technologies. This is on cutting down not only infrastructure costs (CAPEX, Capital Expenditure) but also operational costs (OPEX, Operational Expenditure). Technology for legacy Public Safety networks such as the Project 25 (P.25) or Association of Public Safety Communications Officials (APCO-25), standardized under the Telecommunications Industry Association Engineering Committee TR-8 [2], and the Terrestrial Trunked Radio (TETRA), standardized under ETSI Technical Committee TETRA (TCCE) (TETRA and Critical Communications Evolution) [3], are addressing relatively small markets with few suppliers only and thus do not have the potential to cut down costs to build-out public safety networks and costs of devices (e.g., public safety devices can be very expensive, from $1000 onward, compared to well-known mobile smartphone prices). In contrast, LTE is a global standard, the equipment being able to be used everywhere in the globe with minor adaptations as necessary. This allows reducing production and deployment costs significantly.

On OPEX side, using widely adopted LTE technology allows building public–private partnerships, various studies estimate high savings on total cost of ownership over 10 years of operation. In this scenario, a Public Safety network operator teams up with commercial operators of 3rd Generation Partnership Program (3GPP) networks making use of their infrastructure where possible and potentially enhancing coverage when necessary.

Another driving force is the ability of LTE to provide efficient high speed, low latency, low setup time, and high-security data connectivity, which is the precondition to provide multimedia and especially mission critical multimedia communication. TETRA, for example, provides

LTE for Public Safety, First Edition. Rainer Liebhart, Devaki Chandramouli, Curt Wong and Jürgen Merkel.
© 2015 John Wiley & Sons, Ltd. Published 2015 by John Wiley & Sons, Ltd.

only 7.2 kbit/s data transfer rate per time slot (P.25 offers even less). Four time slots can be combined, which results in a gross data rate of 28.8 kbit/s. New versions of the TETRA standard support data rates up to 691.2 kbit/s.

Another bonus in using LTE is the availability of radio equipment for the most diverse deployment scenarios. LTE base stations are available from macro down to micro, pico, and even femto cell size, that is, at the size of residential Wireless Local Area Network (WLAN) or cable router modems, making it simple to quickly adapt and dynamically set up Public Safety networks even in rural areas where there is no coverage of a macrocellular network.

3.2   What are Public Safety Networks?

In the role model of 3GPP, a Public Safety network is a network providing communication services to public safety entities such as police, firefighters, and civil defense or paramedic services. Depending on the situation in the country of deployment, such networks can also be used by commercially operated security firms, for example, providing security services at airports or at company campuses.

The very first public safety communication networks were based on direct communication between an officer and the dispatcher located in a command center close by. These networks were using specially developed and mostly proprietary technology and special frequency bands for communication. They were isolated solutions, in the worst case not even interoperable across city borders. Another major drawback of this technology was the tight linkage to a single supplier with all the downsides of pricing, unavailability of features, and delivery problems.

As these networks were dedicated to official use only and were strictly governed by a dispatcher located in a command center, they inherently offered priority and preemption like mechanisms by arbitrating communication requests (usually a request to talk) via the dispatcher. The dispatcher was in control of the floor, i.e., deciding who has the right to talk next. Since the mode of communication was mostly "direct communication" mode, these networks also offered intrinsic reliability against network failures, as only the command center and the officer's "walkie-talkie" devices were involved in the communication. Even in case the command center failed, officers were still able to communicate directly to each other. One disadvantage of this direct communication mode was the limited coverage. Soon relay nodes and other network infrastructure equipment were rolled out to extend the communication range, but also adding additional points of failure.

With the upcoming cellular communication systems and the widespread deployment of these networks the interest of public safety authorities to use these technologies was raised, as these networks offer lower deployment costs, multivendor support and with, for example, Global System for Mobile Communications (GSM) growing global coverage.

In parallel to the upcoming second-generation cellular systems, also the development of digital Public Safety communication started in the United States with APCO P.25 and in Europe with TETRA. These systems reduced the fragmentation of the Public Safety communication market quite significantly. P.25 is deployed in more than 50 countries among these are United States, Canada, Australia, New Zealand, Brazil, India, and Russia, while TETRA is used in more than 100 countries, for example, in Europe, Middle East, Africa, Asia Pacific, and Latin America. On a global scale market fragmentation persisted as P.25 and TETRA are not interoperable.

In the mid-1990s of the 20th century a first attempt was made to use GSM for public safety communication. When starting work on TETRA in Europe it was investigated whether to reuse GSM, but it turned out that GSM was missing some important features at that time. Operation on certain frequency bands was not possible and priority/preemption mechanisms were still in their infancy at that time. As a consequence, in Europe, the dedicated cellular TETRA network for public safety communication was standardized, which is nowadays widely adopted. There is one noteworthy exception. Mid of the nineties the GSM technology was enhanced to provide communication services for railways (GSM-R), which have to some extent similar communication needs as public safety authorities. The main reason this succeeded and has been even a success outside of Europe, for example in China, was the radio spectrum dedicated for railways being adjacent to the European GSM 900 frequency band for commercial networks. Thus, no significant changes to the radio system were required and the frequency band was large enough to cope with the traffic demand, without implementation of elaborated priority and preemption mechanisms from the beginning.

With the evolution of GSM from a voice-only to a data-centric network and in parallel moving to a global standardization organization (the 3GPP), the initial problems to make use of 3GPP technology for public safety communication began to disappear. For example, Universal Mobile Telecommunications System (UMTS) and LTE are deployed in a large number of countries around the globe and operating in many different frequency bands. LTE even allows operation on fragmented frequency bands at the same time.

3.3   LTE meets Demands of Public Safety Networks

Besides cost aspects there are many features LTE can already provide for Public Safety use cases without any upgrade. LTE offers interoperability between network equipment of different device and infrastructure manufacturers and between different networks, either national or international (known as roaming), very important in regions such as Europe where networks will likely be operated on a per country basis. Provided roaming agreements between countries exist, users of a Public Safety network in one country will be able to use Public Safety network infrastructure in other countries too. LTE also offers "state-of-the-art" authorization, authentication, and encryption mechanisms ready to be used for Public Safety communication. LTE allows the user to be authenticated toward the network, but also the network to be authenticated toward the user, effectively protecting eavesdropping attacks on Public Safety users.

An intrinsic strength of LTE is the openness of used security standards (for details see 3GPP TS 33.401 [4]) and the large number of security experts monitoring for security breaches and backdoors, providing updates to the standards before these security gaps can be exploited. For example, long time before the first encryption algorithm A5/1 was hacked the new algorithm A5/3 was defined by 3GPP. Owing to the openness of the standards and the global participation from many countries, chances of backdoors implemented in the standards by "interested" parties are less likely to succeed. Even though the parts standardized by 3GPP, especially the communication between device and network, can be considered fairly safe 3GPP networks rely on backhaul and interdomain security where special care has to be taken to prevent attacks.

LTE allows for flexible network sharing mechanisms (see 3GPP TS 23.251 [5]) by which operators, either commercial or Public Safety ones, can share parts of their networks to a certain extent. Most common scenario is Radio Access Network (RAN) sharing where different operators use dedicated core network equipment but share radio equipment with other operators. This allows for reducing deployment costs, but also to overcome gaps in radio coverage.

Priority and pre-emption mechanisms initially designed for disaster scenarios such as tsunamis, earthquakes, and typhoons allow important communication to succeed in case of network overload. These are already in-built features specified for LTE thus they can be easily re-used for Public Safety networks.

LTE provides quality of service on different levels such as per subscriber, per service, or per application. This allows traffic with high requirements on throughput and latency, for example, video streaming, to receive preferential handling over ordinary data transfer.

This has led to the conclusion that only very few features were missing from LTE to allow Public Safety communication. These features were taken up by 3GPP into their work program from Release-12 onwards. The features are concretely: direct mode of communication (for details see Chapter 4 and 3GPP TS 23.303 [6]), group communication (for details see Chapter 5 and 3GPP TS 23.468 [7]), and the Mission Critical Push to Talk (MCPTT) (targeted for 3GPP Release 13, see 3GPP TS 22.179 [8]) service. These additional features can provide a resource efficient group communication mimicking the behavior of "push to talk (PTT)" speech communication of legacy public safety systems. Besides the MCPTT service, other applications can also make use of the enablers specified by 3GPP.

3.4 Wide Range of LTE Devices for Public Safety

As pointed out earlier with the adoption of LTE as radio technology for public safety networks and the implementation of functionalities required to support ProSe or GCSE in the chipsets, it is quite likely that prices for public safety LTE devices will decrease and it also allows various other players to enter the market for public safety devices.

This is especially the case for non- or semi-ruggedized devices where the market already offers semi-ruggedized versions of phones for some time.

Public safety devices are available based on smartphone technology, thus they are able to offer more than just voice, video, group communication, or device-to-device communication, but also applications that enable to streamline the work flow of police officers. Figure 3.1 shows an LTE-based Public Safety tablet from Harris.

LTE will not only change the technical ecosystem but also the economical conditions to bring up interesting new use cases and varieties of devices.

For less harsh operating environments it can be anticipated to use ordinary smartphones or tablets with a few in-built applications to support Public Safety communication. With the exception of mechanical resilience, the boundaries between the highly specialized Public Safety devices and ordinary phones will begin to disappear.

It is expected that Public Safety mobile devices will not only include LTE defined functionality to support Public Safety communication, but also, depending on the region the device is used, TETRA or P.25 functionality to provide service where no LTE coverage is available – a likely scenario during the rollout of LTE Public Safety systems. The interworking between these two applications is out of scope of the 3GPP work and is left to the individual device manufacturer and national/regional regulation.

Figure 3.1   Harris RF-3590 LTE Public Safety Tablet

3.5   Standalone versus Shared Deployments

In most of the cases, Public Safety networks have been designed as standalone networks, usually deployed by national authorities or parts of the government. By doing so it was rather
simple to be in full control of the network assuring privacy and secrecy. Deploying a network
this way comes with high costs for providing a high quality, high reliable, for being fully
government controlled network in the end.

As expected, this is the most expensive way of providing Public Safety networks, not being
able to fully assess the installation costs in the beginning. Thus, in many countries the initial
cost assessments for the final deployment of legacy public safety systems with coverage in all
relevant areas were exceeded significantly in the end.

To overcome such a situation, commercial mobile operators were also confronted with
3GPP developed RAN sharing mechanisms allowing operators to share radio base stations
and implicitly also sites or part of sites where base stations are physically deployed. If a
Public Safety network operator has a sharing agreement with national mobile operators he
benefits from the combined coverage and capacity of these networks, not only in areas where
one network may have fringe coverage but also in areas where peak traffic is to be expected
such as densely populated city centers. With (national) roaming agreements in place the

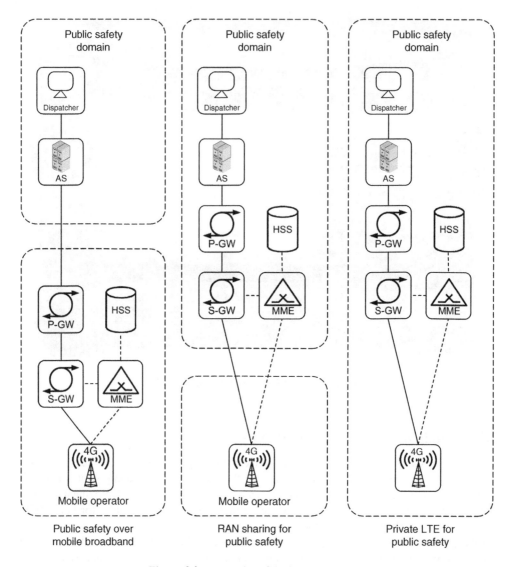

Figure 3.2 Examples of deployment options

Public Safety network operator could even make use of the infrastructure of commercial operators in certain areas without the need to deploy its own infrastructure. Figure 3.2 shows the examples of deployment options, ranging from a fully mobile network operator-hosted deployment scenario to a dedicated, standalone Public Safety deployment scenario.

Another benefit comes with the higher network/service availability and reliability when using more than one network. It reduces the probability of outages caused by technical problems by the number of operators involved, provided these operators all have an independent infrastructure in the area.

3.6 Interworking

3.6.1 Device Aspects

It is expected that an LTE-based public safety device (in terms of 3GPP the UE) will also provide legacy network support, for example, for TETRA or P.25. The working assumption is that the decision which of the two networks to use will be controlled by the application running on the UE. Even if 3GPP is going to specify the MCPTT application, it does not have a mandate to work on topics outside of its scope, that is, the 3GPP-standardized MCPTT application will not specify in detail the interworking to P.25 or TETRA systems.

But of course, the 3GPP standardized part will not be completely agnostic to the fact that there are other ways of communication for a public safety device. For example, to enable decisions when to switch over from network-based GCSE to ProSe-based communication necessary information from the lower 3GPP layers can be provided to the MCPTT application. Parts of this information in conjunction with information from the legacy part of the device can also be used to decide when to switch from LTE to P.25/TETRA and vice versa.

3.6.2 Network Aspects

The other interworking scenario is when a device currently connected via LTE and using MCPTT wants to communicate to one or several devices of which some might only be reachable via legacy public safety networks or vice versa. In this case the network has to ensure interworking on the different layers, for example, transport and application layer. For the application layer this might require protocol translation and transcoding capabilities in case devices are using different codecs.

References

[1] Global Mobile Suppliers Association (GSA): http://www.gsacom.com/.
[2] Telecommunications Industry Association (TIA) TR-8: http://www.tiaonline.org/all-standards/committees/tr-8.
[3] ETSI TCCE: http://www.etsi.org/technologies-clusters/technologies/tetra.
[4] 3GPP TS 33.401: "3GPP System Architecture Evolution (SAE); Security Architecture".
[5] 3GPP TS 23.251: "Network Sharing; Architecture and Functional Description".
[6] 3GPP TS 23.303: "Proximity-Based Services (ProSe)".
[7] 3GPP TS 23.468: "Group Communication System Enablers for LTE (GCSE_LTE)".
[8] 3GPP TS 22.179: "Mission Critical Push to Talk MCPTT (Release 13)".

4

Proximity Services

4.1 Introduction to Proximity Services

4.1.1 Proximity Services Overview

Proximity Services, in short ProSe, provides mechanisms for devices to discover other devices in close proximity and to communicate with other devices directly, that is, without the data path being routed via the network infrastructure. This is illustrated with an example – Fireman John and Fireman Bob are located in the same building and they are trying to discover each other's presence and communicate when they are in proximity. In a traditional cellular network, if John and Bob want to communicate even when they are in proximity, their devices have to connect to the network first in order to transmit any kind of data through the mobile radio and core network infrastructure. With the introduction of ProSe, devices like the ones owned by John and Bob can discover each other and transmit user plane traffic directly between them in proximity. The "proximity range" depends certainly on the strength of the radio signal and other radio conditions such as interference. The actual range (e.g., 50 m, 200 m, or 1 km) varies depending on the power level used for transmitting the radio signal.

The ProSe function consists mainly of mechanisms on LTE radio level in order to enable direct discovery and communication between two or more devices. Functions and capabilities introduced by ProSe can be used by any application running on top of a ProSe-enabled device. Public Safety (PS) is one of the ProSe use cases, file sharing and friend finder are potentially other use cases.

ProSe partly breaks the paradigm for the User Equipment (UE) to "receive before transmit" and complicates radio interference situations by UEs now transmitting on frequencies they previously only used to listen. ProSe is an optional functionality and it is up to the manufacturer of a UE to decide whether to implement the ProSe feature. There are two types of ProSe-enabled UEs, one supporting "social" or "commercial" use cases for ordinary mobile subscribers, referred to as "ProSe-enabled non-Public Safety UE" and the other supporting "PS" use cases, referred to as the "ProSe-enabled Public Safety UE" for use by PS personnel such as first responders, and police fire brigades. They differ in the ProSe-related feature set they have to support. For PS use cases, direct communication is a must have feature, while direct discovery can be omitted.

LTE for Public Safety, First Edition. Rainer Liebhart, Devaki Chandramouli, Curt Wong and Jürgen Merkel.
© 2015 John Wiley & Sons, Ltd. Published 2015 by John Wiley & Sons, Ltd.

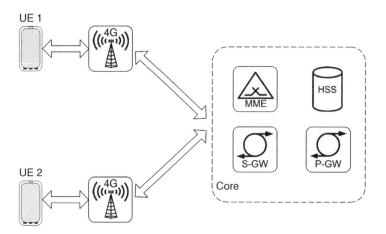

Figure 4.1   Data paths without using ProSe

4.1.2   ProSe Communication

Before the introduction of PS all data traffic originating or terminating from/to a UE had to pass through at least parts of the network as depicted in Figure 4.1.

There had already been attempts in the past to define direct communication between UEs. In the late 1990s, a direct mode of operation was defined for Global System for Mobile Communications (GSM)/GSM/EDGE Radio Access Network (GERAN)-based railway communication but has not been deployed due to radio interference problems and consequently the lack of chipsets supporting it.

With the advent of more modern radio technologies and the surge in mobile data communication caused by Smart phones, the situation began to change. Operators were confronted with a tremendous increase in data traffic per device and, at the same time, decrease in revenue per amount of data transferred. To cope with this, 3rd Generation Partnership Program (3GPP) had already specified local breakout mechanisms such as Local Internet Protocol Access (LIPA) or Selective IP Traffic Offload (SIPTO). These mechanisms allow low-revenue traffic to exit to local area networks at home or a campus, or to connect to the Internet close to the user before consuming too much network resources.

As some of this low-revenue traffic involves only sender and receiver that are close together, for example, friends sitting together sharing pictures or playing online games, the idea for direct communication was becoming obvious again.

This communication scenario in Figure 4.2 is referred to as "ProSe communication" or "Direct Communication." ProSe Communication can take place using Long-Term Evolution (LTE) or Wireless Local Area Network (WLAN) radio technology. In case of WLAN radio technology, ProSe only assists in the connection establishment and with service continuity aspects, as the WLAN radio technology is out of scope of 3GPP and allows already for a direct communication mode ("WLAN *ad hoc* networks"). The mobile network can control resources used for ProSe communication and must authorize the use of ProSe communication. The most obvious reason for this being the use of licensed spectrum the operator owns,

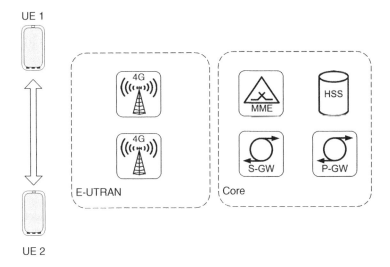

Figure 4.2   The direct communication path

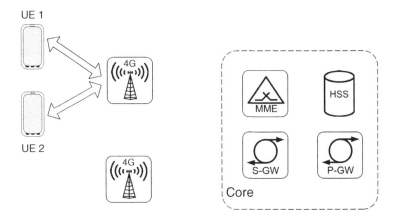

Figure 4.3   The locally routed data path

further, in many countries, by licensing terms, operators are obliged to be in control how their licensed spectrum is used.

To have even stricter means of resource usage control, a locally routed communication scenario was introduced by 3GPP by which traffic is routed via an Evolved NodeB (eNodeB) in reach of both communication partners (Figure 4.3).

This mode of operation, however, defeats the purpose of efficient resource usage as the licensed spectrum is expensive and the potentially congested radio links to the eNodeB are still being used. For this reason, the locally routed data path scenario is not expected to be widely deployed and was not specified in Release 12.

4.1.3   ProSe Discovery

Motivation for ProSe Discovery is mainly the desire to provide new proximity services for social applications, advertisement, and location-based services to mobile devices. Nowadays, location-based services mainly rely on a combination of Global Positioning System (GPS), cellular (e.g., Assisted GPS), and Wi-Fi technologies.

ProSe Discovery is a competing feature allowing a UE to discover that another UE is in proximity. ProSe Discovery can be considered a stand-alone feature; its use is not necessarily tied to or followed by ProSe communication. As a result, ProSe Discovery is a functionality that can be offered as a stand-alone service to ProSe-enabled UE(s). However, in order to communicate with each other directly, Public Safety communication partners in proximity would benefit of previous ProSe Discovery. The authorization to use the discovery mechanism can be given to a UE in general or only to certain applications residing on the UE. For example, the operator could allow the use of ProSe Discovery mechanisms for a "friend finder" application but not for locating shops or for updating the contacts list of a chat application.

The ProSe Discovery feature is defined in an "open mode" and in a "restricted mode." For the open mode, no permission is needed from the UE that is to be discovered, while the restricted mode requires permission from the UE that is being discovered.

The criteria to decide when a UE is assumed to be in range of another UE (in the context of ProSe) can be defined by the operator and likewise the operator has to authorize the use of the discovery mechanism and resources by a UE.

Following are the two different discovery procedures supported in Release 12:

1. **ProSe Direct Discovery**
2. **EPC-level Discovery**

The most interesting discovery mechanism for Public Safety devices is "ProSe Direct discovery" or simply referred to as "Direct discovery" as this enables the UE to discover other UEs without network coverage. Public Safety use cases may not require an automated discovery procedure in the device as Public Safety personnel may recognize each other within a certain area (e.g., if two firemen are in direct in visibility). However, in isolated firefighter scenarios, direct discovery is an essential feature for Public Safety use cases.

Also, if a police officer wishes to explicitly disconnect from the network, Evolved Packet Core (EPC) (network) assisted discovery or EPC-level discovery cannot be used. EPC-level discovery is expected to be used mostly for commercial reasons and in cases where the devices or the network do not support the direct discovery mode. If two friends are entering a football stadium and trying to discover each other's presence, their devices can use the EPC-level discovery mechanism to alert each other when they are in close proximity.

4.1.4   ProSe for Public Safety

The mechanisms described earlier constituting the ProSe feature can also be used for PS communication purposes. One core requirement for PS scenarios is a direct link communication between two or more UEs. The main reason for this requirement is to have a fallback communication solution in case there is a network outage or the users themselves decide to disconnect from the network and use direct communication.

However, ProSe has to fulfill some special requirements when it is being used for PS communication. This is why 3GPP defined a special type of ProSe-enabled UE, namely, the "ProSe-enabled Public Safety UE".

In contrast to an ordinary ProSe-enabled UE, the Public Safety ProSe-enabled UE may initiate ProSe communication without previous discovery to accelerate the setup of communication, for example, in emergency situations. This use case most closely resembles the "walkie-talkie" style of communication between legacy Public Safety communication devices, in which the reception of the communication by others cannot be guaranteed.

Unlike commercial communication scenarios, the Public Safety ProSe-enabled UE can establish ProSe Direct communication links directly with other Public Safety ProSe-enabled UEs, regardless of whether the Public Safety ProSe-enabled UE is served by Evolved Universal Terrestrial Radio Access Network (E-UTRAN) or not. A Public Safety ProSe-enabled UE is always capable to participate in ProSe Direct communication between two or more PS ProSe-enabled UEs which are in proximity to each other.

A Public Safety ProSe-enabled UE still needs to be authorized by the network operator to make use of ProSe. Requesting authorization from the network to use ProSe may not work in all cases as the Public Safety ProSe-enabled UE can also be out of network coverage or other UEs that are out-of-coverage require direct communication. As a consequence, this kind of authorization has to be provided either by configuring the UE or Universal Subscriber Identity Module (USIM) beforehand. There is even a requirement for a new UE to work "out of the box" when powered up for the first time while being without network coverage, the use case for this being, for example, a firefighter having to replace his UE in an emergency situation without network coverage.

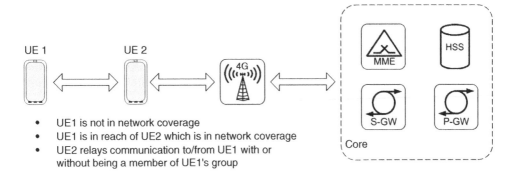

- UE1 is not in network coverage
- UE1 is in reach of UE2 which is in network coverage
- UE2 relays communication to/from UE1 with or without being a member of UE1's group

- UE1 is not in reach of UE3
- UE2 is in reach of UE1 and UE3 and relays the communication with or without being a member of the group of UE1 and/or UE3

Figure 4.4   Public Safety relay scenarios using ProSe

There are two ProSe functions that are only available to Public Safety ProSe-enabled UEs. The first one being the ProSe UE-to-Network Relay where one Public Safety ProSe-enabled UE that is in network coverage relays communication between the network and other Public Safety ProSe-enabled UEs that might not be in network coverage. The second one being the ProSe UE-to-UE Relay by which a Public Safety ProSe-enabled UE relays ProSe communication to other Public Safety Prose-enabled UEs that might not be in close proximity to each other. Both functions are intended, for example, for scenarios where firemen entering a building lose network coverage but still are in contact with their commander or car outside the building. Figure 4.4 shows the described relay scenarios.

In the subsequent sections, we describe how proximity services can actually be realized in LTE networks and LTE devices.

The stage 1 feasibility study on proximity services was documented in 3GPP TR 22.803 [1] while the majority of service requirements were documented in 3GPP TS 22.278 [2].

4.2   Proximity Services Architectures

The main new functional entity introduced in the 3GPP architecture in TS 23.303 [3] to support Proximity Services is the so-called ProSe Function. This functional entity plays different roles for each feature supported in ProSe. The ProSe Function can be present in the Home Public Land Mobile Network (HPLMN), Visited Public Land Mobile Network (VPLMN), and Local PLMN. Please see the "Terms and Definitions" section for definitions of HPLMN, VPLMN, and Local PLMN.

In addition, a ProSe Application Server was introduced to manage PS (mainly discovery and communication) at the application layer. For the following architecture descriptions, we assume that there are two UEs (UE-A and UE-B) trying to communicate to each other using Proximity Services functionality (i.e., UE-A perform direct discovery and direct communication with UE-B).

The new reference points introduced for ProSe had to accommodate the new ProSe Function, ProSe Application Server, and enable direct discovery, EPC-level discovery, and the direct communication path between two UEs.

4.2.1   Non-roaming Architecture

This architectural model (see Figure 4.5) assumes the simplest configuration in which two UEs (UE-A and UE-B) trying to communicate with each other, have subscribed to the same PLMN and both UEs are registered in their home PLMN.

Example scenario: Fireman Bob and Fireman John are subscribed to the same Public Safety network PLMN A. Bob and John are in coverage of their home network located in the same country. Bob and John are trying to communicate with each other.

An overview of the main ProSe reference points, their roles, and the underlying protocols can be found in Section 4.2.5.

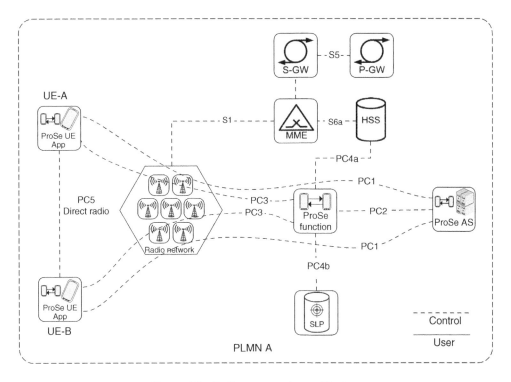

Figure 4.5   ProSe non-roaming architecture

4.2.2   Inter-PLMN Architecture

This architectural model (see Figure 4.6) assumes the two UE(s) UE-A and UE-B have sub-scribed to different PLMN(s) PLMN A and PLMN B but the PLMN(s) are located in the same country. It also assumes that the two UE(s) are registered in their home PLMNs.

Example scenario: Fireman Bob is subscribed to Public Safety network operator A and fire-man John PS to operator B. Operator A and Operator B offer services in the same country. Bob and John are located in the same country. Bob and John are trying to communicate with each other directly.

4.2.3   Roaming Architecture

This configuration is more complex. It is assumed that the UE(s) have subscribed to differ-ent PLMN(s) and one of the UEs is roaming, for example, UE-A is subscribed to PLMN A but is roaming in PLMN C. UE-B is subscribed to PLMN B and located in PLMN B. This architectural model is shown in Figure 4.7.

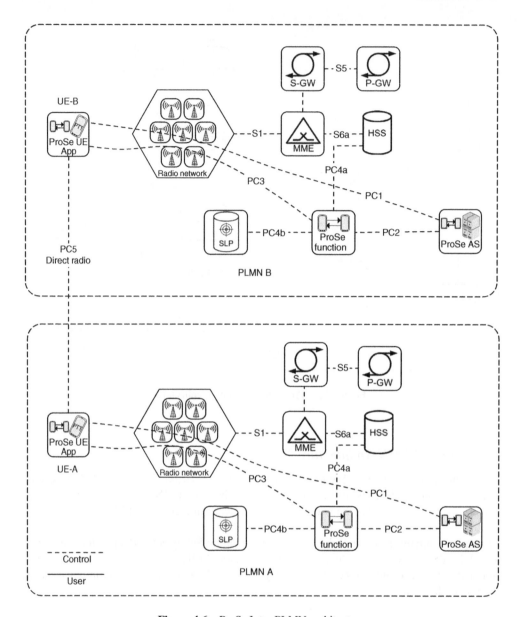

Figure 4.6 ProSe Inter PLMN architecture

Example scenario: Fireman Bob is subscribed to Public Safety network operator A, but is roaming in commercial network C (due to lack of coverage of network A). Fireman John is subscribed to Public Safety network B and located in the same Public Safety network B. Bob and John are trying to communicate with each other directly.

4.2.4   Description of Functional Entities

This section provides a description of functional entities that are required in order to support ProSe. These functions are needed in addition to the basic functions required for LTE/SAE as described in chapter 1. Functional entities that are not impacted by ProSe are not described in this section.

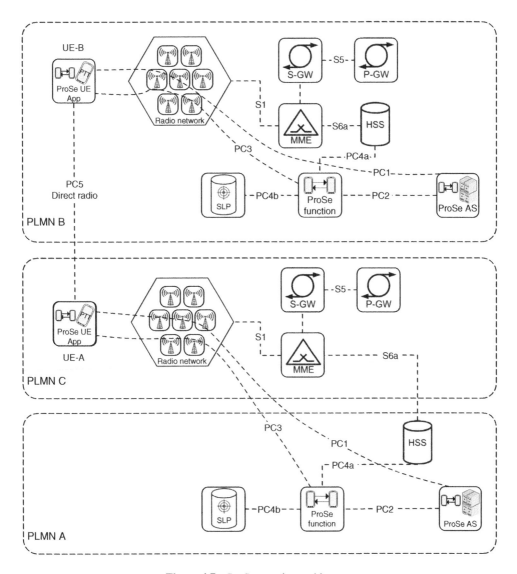

Figure 4.7   ProSe roaming architecture

4.2.4.1   ProSe Function

This newly introduced ProSe Function plays different roles in order to support all the features required for ProSe. In 3GPP Release 12, it has been assumed that there is only one ProSe Function in each PLMN supporting ProSe. Thus, if an operator deploys multiple ProSe Functions within the same PLMN, there is no standardized procedure for the UE to discover all of these ProSe Functions (the simplest way is certainly to configure a logical name or address of the ProSe Function on the devices). In real deployments, it is possible that this ProSe Function is not implemented as a separate network element but is collocated with another network element that resides in the user plane path of the UE. This would allow that an already deployed network element is just upgraded to support the functionalities required for the ProSe Function.

The ProSe Function consists of three main subfunctions that perform different roles depending on the ProSe feature:

- The **Direct Provisioning Function (DPF)** is used to provision the UE with necessary parameters in order to use ProSe Direct Discovery and ProSe Direct Communication. The UEs are provisioned with PLMN specific parameters allowing them to use ProSe in a specific PLMN. This includes information such as list of PLMNs in which a UE can perform Direct Discovery and parameters needed for Direct Communication when the UE is "out of network" coverage.
- The **Direct Discovery Name Management Function** is used for open ProSe Direct Discovery (also referred to as ProSe Direct Discovery in an "open mode") to allocate and process the mapping of ProSe Application IDs and ProSe Application Codes. It uses ProSe-related subscriber data stored in the Home Subscriber Server (HSS) for authorization of each discovery request. It also provides the UE with necessary security data to protect discovery messages exchanged over the air interface.
- The **EPC-level Discovery ProSe Function** is used to provide network-assisted discovery using location information to UE(s).

In addition, the ProSe Function provides the necessary functionality to collect charging data for usage of ProSe Direct Discovery, ProSe Direct Communication, and EPC-level ProSe Discovery in HPLMN, VPLMN, and Local PLMN.

The IP address of a ProSe Function can be discovered through interactions with Domain Name System (DNS). The Fully Qualified Domain Name (FQDN) of a ProSe Function in the HPLMN may either be preconfigured in the UE or automatically provisioned by the network or self-constructed by the UE, for example, derived from the HPLMN Identity using a certain scheme.

Figure 4.8 shows the internal ProSe Function structure.

Figure 4.9 provides an overview of the different ProSe Function interfaces.

4.2.4.2   ProSe Proxy Function

The ProSe Proxy Function is located in the VPLMN when a UE is roaming and local breakout is applied, that is, the Packet Data Network (PDN) connection is terminated at a Packet Data Network Gateway (P-GW) in the VPLMN. In such a use case it allows communication between

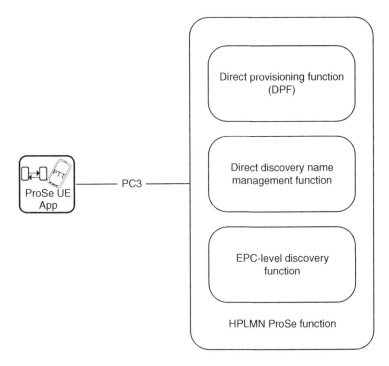

Figure 4.8 ProSe Function internal structure

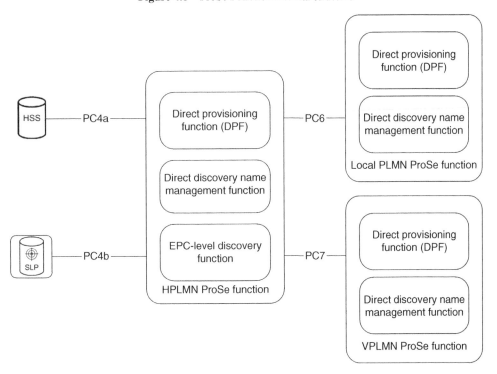

Figure 4.9 ProSe Function interfaces

a UE in VPLMN and the ProSe Function in HPLMN. Such proxy functions might be needed when signaling connection between UE and ProSe Function in HPLMN are assumed to be insecure. The UE is not aware that a ProSe Proxy Function is present.

4.2.4.3 User Equipment

"ProSe-enabled UEs" can be, in principle, classified into "ProSe-enabled Public Safety UEs" and "ProSe-enabled non-Public Safety UEs" (i.e., ProSe-enabled commercial UEs).

In addition to the functions described in chapter 1 for an LTE capable UE, any ProSe-enabled UE may support the following functions:

– Exchange of control information with a ProSe Function like request for service authorization and discovery request for announcing (see Section 4.5.1).
– Open ProSe Direct Discovery of other ProSe-enabled UEs over the PC5 reference point.

In addition to the above-mentioned functions, ProSe-enabled Public Safety UEs can support the following functions:

– One-to-many ProSe Direct Communication over PC5 reference point.
– Act as a ProSe UE-to-Network Relay (the relay feature is not specified in Release 12). The Remote UE (the UE without network coverage) communicates with the ProSe UE-to-Network Relay over the PC5 reference point and gets access to network services.
– Exchange of control information between ProSe UEs over PC5 reference point, for example, for UE-to-Network Relay detection and ProSe Direct Discovery.
– Exchange of ProSe control information between another ProSe-enabled UE and the ProSe Function over PC3 reference point. In the ProSe UE-to-Network Relay case, the Remote UE will send this control information over PC5 user plane to be relayed over the LTE-Uu interface toward the ProSe Function.
– Configuration of parameters including UE's IP address, ProSe Layer-2 Group IDs, Group security data, and radio resource parameters. These parameters can be configured in the UE, or, if in-coverage, provisioned by the ProSe Function over PC3.

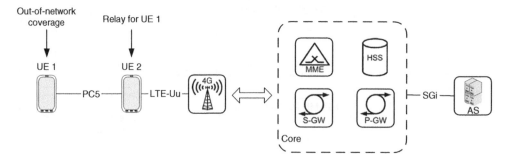

Figure 4.10   ProSe UE-to-Network Relay

4.2.4.4   ProSe UE-to-Network Relay

The ProSe UE-to-Network Relay (see Figure 4.10) provides connectivity to "unicast" services for Remote UEs that are not in network coverage. Although this function was introduced to the ProSe architecture, it was not specified in 3GPP Release 12.

The ProSe UE-to-Network Relay can relay unicast traffic (in Uplink (UL) and Downlink (DL)) between the Remote UE and the network. The ProSe UE-to-Network Relay can also provide generic functions to relay any type of traffic that is relevant for Public Safety communication. If the Remote UE moves out of ProSe UE-to-Network relay coverage, then its IP address is not preserved.

4.2.4.5   ProSe Application Server

The ProSe Application Server mainly supports application layer functions for EPC-level discovery. It stores data such as the EPC ProSe user ID (EPUID) and ProSe Function ID (PFID). It also supports mapping of Application Layer user ID (ALUID) and EPUID.

4.2.4.6   MME

In addition to functions described in Chapter 1, the MME is assumed to support the following functions for ProSe:

– Receive ProSe-related subscription information from the HSS.
– Obtain UE capability data regarding the direct discovery feature.
– If the UE supports the necessary capability and the UE is authorized to use the corresponding service, then provide an indication to the eNB over S1-AP signaling (see 3GPP TS 36.413 [10]) that the UE is authorized to use ProSe services.

4.2.4.7   SUPL Location Platform (SLP)

SLP provides location information to the ProSe function for EPC-level discovery procedures.

4.2.5   Interfaces and Protocols

4.2.5.1   Control Plane

The control plane consists of protocols to control and support the establishment, modification, and termination of the user plane. This includes the following functions:

– Control the UE configuration data related to ProSe.
– Control of ProSe Direct Discovery functions.
– Control the setup of the connection between Remote UE and ProSe UE-to-Network Relay.
– Control attributes of an established network connection, such as activation of an IP address.

UE–ProSe Function
ProSe control plane signaling between UE and ProSe Function (PC3 interface) is carried over the user (data) plane (see Figure 4.11).

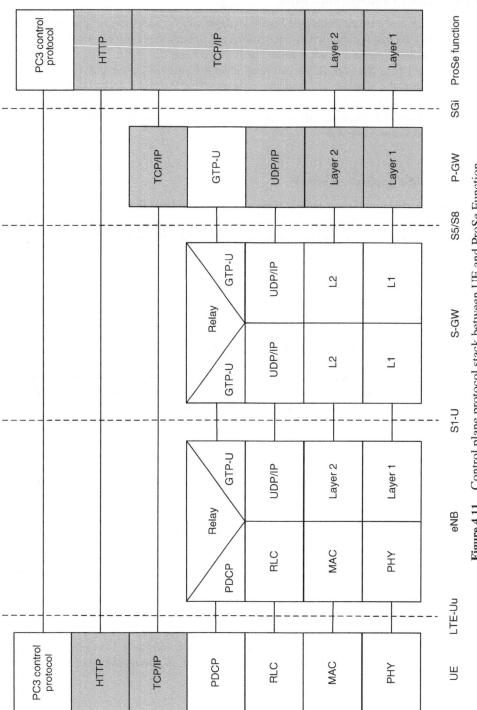

Figure 4.11   Control plane protocol stack between UE and ProSe Function

PC3 Control Protocol
It is used to enable communication between a ProSe-enabled UE and the ProSe Function and is specified in 3GPP TS 24.334 [5]. The UE and ProSe Function use the Hypertext Transfer Protocol Version 1.1 (HTTP 1.1) as specified in IETF RFC 2616 [11] as transport protocol for ProSe control plane messages over PC3.

HSS–ProSe Function
Figure 4.12 shows the control plane protocol stack between ProSe Function and HSS.

PC4a DIAMETER Application
Main function of PC4a is the transfer of subscription and authentication data between HSS and ProSe Function. DIAMETER is defined in IETF RFC 3588 [12] and the PC4a DIAMETER application in 3GPP TS 29.344 [6].

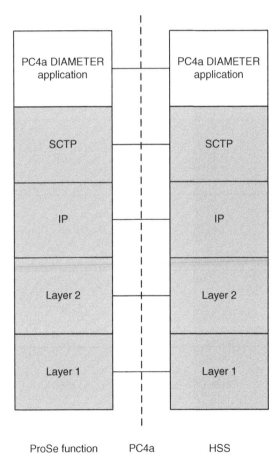

Figure 4.12   Control plane protocol stack between ProSe Function and HSS

SLP–ProSe Function

Figure 4.13 shows the control plane protocol stack between ProSe Function and SUPL Location Platform (SLP).

Mobile Location Protocol (MLP)

It is used to carry location information as specified in Open Mobile Alliance Location Interoperability Forum Mobile Location Protocol (OMA LIF MLP) between SLP and ProSe Function.

UE–UE

Figure 4.14 shows the control plane protocol stack between two ProSe-enabled UEs.

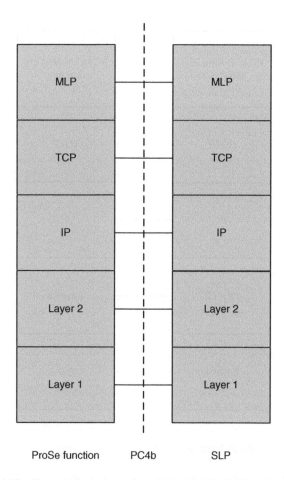

Figure 4.13   Control plane protocol stack between ProSe Function and SLP

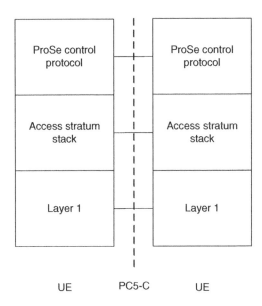

Figure 4.14   Control plane protocol stack between two ProSe-enabled UEs

ProSe Control Protocol
It is used for handling control messages to support ProSe Direct Discovery and ProSe UE-to-Network Relay Discovery and is specified in 3GPP TS 24.334 [5].

Access Stratum
Access Stratum (AS) performs the following functions:

- Interfaces with Upper Layer: The Medium Access Control (MAC) layer receives the discovery information from the upper layer (i.e., upper layer refers to application layer or Non Access Stratum (NAS) layer). The IP layer is not used for transmitting the discovery information. The discovery information is transparent to Access Stratum.
- Scheduling: The MAC layer determines the radio resource to be used for transmitting the discovery information.
- Discovery Protocol Data Unit (PDU) generation: The MAC layer builds the MAC PDU carrying the discovery information and sends the MAC PDU to the PHY layer for transmission over the radio resource.

ProSe Function–ProSe Function
Figure 4.15 shows the control plane protocol stack between two ProSe functions. PC6 is an inter-PLMN interface and PC7 is the corresponding roaming interface.

PC6/PC7 DIAMETER Application
It supports the transfer of subscriber location related information between ProSe Functions over the PC6 and PC7 interfaces. This interface is used when the ProSe Function in the

HPLMN collects authorization and configuration information from other PLMNs. In roaming scenarios it is also used to authorize ProSe Direct Discovery requests, retrieve the Discovery Filter(s) corresponding ProSe Application ID name(s), and translate the ProSe Application Code to the ProSe Application ID Name. Further details regarding PC6/PC7 interfaces can be found in 3GPP TS 29.345 [9].

ProSe Function–ProSe Application Server
Figure 4.16 shows the control plane protocol stack between ProSe Function and ProSe Application Server.

PC2-AP
PC2-AP is the PC2 Application Protocol and is specified in 3GPP TS 29.343 [4]. This interface is used for EPC-level ProSe Discovery, for functions such as create the mapping between application level user identifiers and EPC ProSe User IDs.

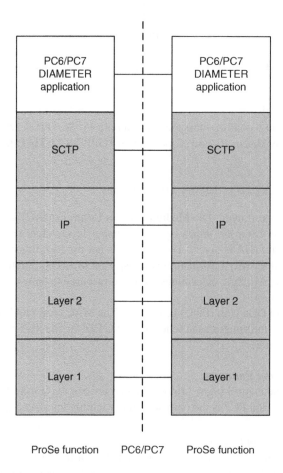

Figure 4.15   Control plane protocol stack between ProSe Functions

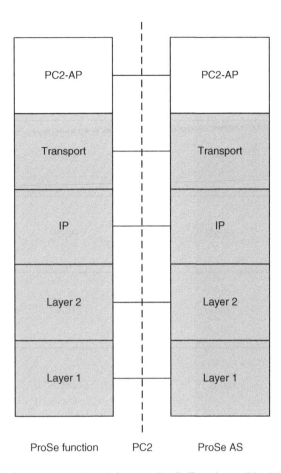

ProSe function   PC2   ProSe AS

Figure 4.16   Control plane protocol stack between ProSe Function and ProSe Application Server

User Plane
UE–UE
Figure 4.17 shows the user plane protocol stack between two ProSe-enabled UEs.

PC5-U
The E-UTRA radio protocols PHY, MAC, Radio Link Control (RLC), and Packet Data Convergence Protocol (PDCP) between two UEs (similar to the ones used between eNB and UE) are described in Chapter 1 and specified in 3GPP TS 36.300 [8].

UE–UE-to-Network Relay
Figure 4.18 shows the user plane protocol stack between a ProSe-enabled UE and a UE-Network-Relay.

Summary of Reference Points and Protocols
Table 4.1 summarizes the reference points and protocols used for ProSe.

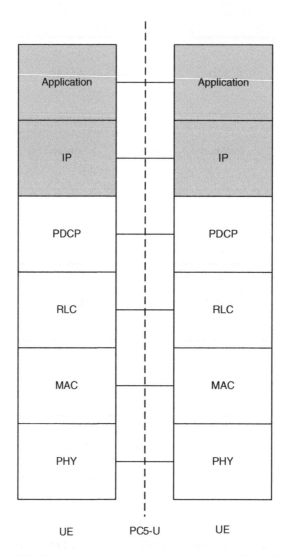

Figure 4.17   User plane protocol stack between two ProSe-enabled UEs

4.3   Synchronization

The term "synchronization" means, in general, coordination of events to operate a system in unison. Just as the term goes, synchronization is important for a device to successfully camp in a cellular network and for a device to discover and communicate with other devices directly. This section provides an overview of so-called LTE primary and secondary synchronization signals that are required for cell synchronization and Device-to-Device (D2D) synchronization signals that are required for D2D discovery and D2D communication.

   We use the term "D2D" in this context as a synonym for "ProSe."

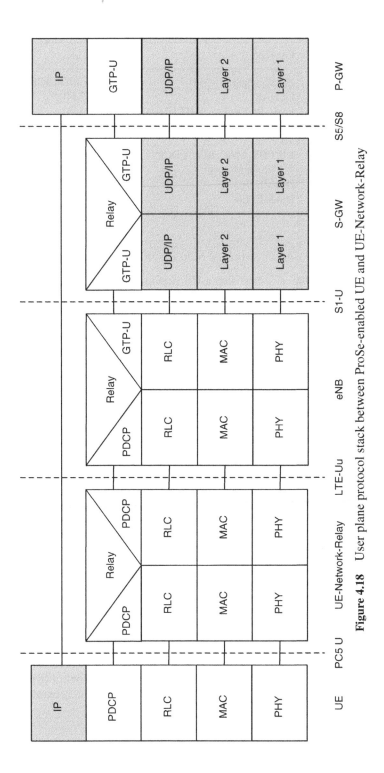

Figure 4.18 User plane protocol stack between ProSe-enabled UE and UE-Network-Relay

Table 4.1 Summary of ProSe reference points and protocols

Reference point	Protocols	Specifications
PC1	Not specified in 3GPP	Not specified in 3GPP
PC2	PC2-AP	TS 29.343 [4]
PC3	ProSe Protocol (over HTTP)	TS 24.334 [5]
PC4a	DIAMETER	TS 29.344 [6]
PC4b	MLP	OMA LIF MLP [7]
PC5	UP: PHY, MAC, RLC, PDCP	TS 36.300 [8], TS 24.334 [5]
	CP: ProSe Protocol	
PC6	DIAMETER	TS 29.345 [9]
PC7	DIAMETER	TS 29.345 [9]

4.3.1 LTE Primary and Secondary Synchronization Signals

Cell synchronization is the very first step a UE performs when it wants to camp on a cell. During synchronization the UE acquires a Physical Cell ID (PCI), time slot, and frame synchronization, which will enable the UE to read System Information Blocks (SIBs) from a particular network (see also 3GPP TS 36.300 [8]).

The UE will tune its radio by tuning to different frequency channels depending on the bands it is supporting. Assuming that it is currently tuned to a specific band or channel, the UE first finds the Primary Synchronization Signal (PSS), which is located in the last Orthogonal Frequency Division Multiplexing (OFDM) symbol of the first time slot of the first subframe (subframe 0) of a radio frame. This enables the UE to be synchronized on subframe level. The PSS is repeated in subframe 5, which means the UE is synchronized on a 5 ms basis because each subframe is 1 ms long. From the PSS, UE is also able to obtain physical layer identity (0 to 2).

In the next step, the UE finds the Secondary Synchronization Signal (SSS). SSS symbols are also located in the same subframe of PSS but in the symbol before PSS as shown in Figure 4.19. From SSS, the UE is able to obtain the physical layer cell identity group number (0 to 167).

Using physical layer identity and cell identity group number, UE knows the PCI for this cell. In LTE 504 physical layer cell identities are allowed and are divided into unique 168-cell layer identity groups where each group consists of three physical layer identities. As mentioned earlier, the UE detects physical layer identity from PSS and physical layer cell identity group from SSS and calculates the PCI.

Once the UE knows the PCI for a given cell, it also knows the location of cell reference signals. Reference signals are used in channel estimation, cell selection, cell reselection, and handover procedures.

After the cell synchronization procedure (reference 3GPP TS 36.211 [13]), the UE will proceed to read the Master Information Block (MIB) and other SIBs like SIB1 and SIB2 (refer to Terms and Definitions section for definitions of MIB and SIB). Once it has the necessary parameters, the UE can request the eNodeB for a Radio Resource Control (RRC) connection establishment so that it can exchange signaling information with the network. Then, it uses

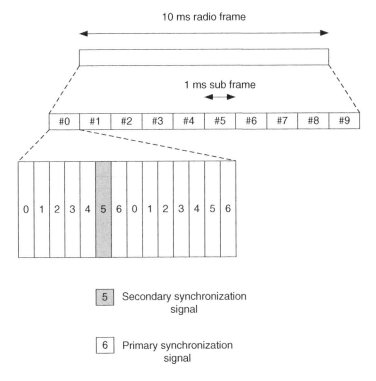

Figure 4.19 LTE Synchronization Signal

the established RRC connection to register with the network and this process is called "Initial Attach" (explained in Section 1.6.9).

4.3.2 LTE D2D Synchronization

A D2D Synchronization Source transmits at least a D2D Synchronization Signal (D2DSS). The D2D Synchronization Source can either be a ProSe-enabled UE or an eNodeB. The transmitted D2DSS is used by a UE to obtain time and frequency synchronization information. The D2DSS transmitted by a D2D Synchronization Source, which is an eNodeB when the UE is in network coverage, is the well-known Release 8 LTE PSS/SSS. The structure of D2DSS transmitted by D2D Synchronization Sources other than the eNodeB follows the signal design defined for D2D in Release 12. Working assumption is that a synchronization source has a physical layer identity known as Physical Layer Identity (PSSID). At the time of writing this book, the structure of D2DSS is yet to be finalized in 3GPP RAN Working Group 1.

The concept of a Physical D2D Synchronization Channel (PD2DSCH) transmitted by a D2D Synchronization Source is likely to be defined in 3GPP.

If a D2D Synchronization Source is detected, the UE synchronizes its receiver to the source before it transmits the D2DSS. UEs may transmit at least a D2DSS derived from the D2DSS received from a D2D Synchronization Source (e.g., another ProSe-enabled UE). If a UE transmits a D2DSS, certain rules are followed for determining which D2D Synchronization Source the UE uses as the timing reference for its transmissions of D2DSS. Rules for determining which D2D Synchronization Source the UE uses as timing reference for its transmissions of D2D signal are as follows:

- D2D Synchronization Sources, which are eNodeBs, have a higher priority than D2D Synchronization Sources which are UEs
- D2D Synchronization Sources, which are UEs in-network-coverage, have a higher priority than D2D Synchronization Sources, which are UEs out-of-network-coverage
- After giving priority to D2D Synchronization Sources, which are eNodeBs, followed by UEs in-network-coverage, selection of D2D Synchronization Source is based on metrics such as received D2DSS quality.

If no D2D Synchronization Source is detected, a UE may nevertheless transmit D2DSS. A UE may reselect the D2D Synchronization Source it uses as the timing reference for its transmissions of D2DSS, if the UE detects a change in the D2D Synchronization Source(s). Detailed rules for selection are for further study in 3GPP.

It is assumed that D2DSS transmission configuration is the same for ProSe Direct Discovery and ProSe Direct Communication, if the network supports both. D2D synchronization is a prerequisite to perform ProSe Direct Discovery and ProSe Direct Communication.

4.4   Service Authorization

Upon successful synchronization and registration with the network, a ProSe-capable UE needs to obtain (service) authorization for use of ProSe Direct Discovery or ProSe Direct Communication or both. This is usually valid for certain duration (validity time). The request for authorization is normally triggered due to one of the following events:

1. In order to initiate ProSe Direct Discovery or ProSe Direct Communication.
2. The UE moves to a new registered PLMN when it is already engaged in either ProSe Direct Discovery or ProSe Direct Communication.
3. When the validity time for the service authorization expires.

In case of both roaming and non-roaming scenarios, the UE should obtain service authorization from the ProSe Function in the HPLMN as shown in Figure 4.20.

The authorization is happening using IP-based mechanisms and only IP connectivity is required for the UE to contact the ProSe Function.

In case of roaming scenarios, the ProSe Function in the HPLMN obtains authorization from the ProSe Function in the VPLMN.

If the UE can monitor for other UEs in proximity in PLMNs other than the serving PLMN (in the so-called Local PLMNs), then the UE may need to be authorized by the Local PLMN and the ProSe Function in the HPLMN obtains authorization information from the ProSe Function in the Local PLMN.

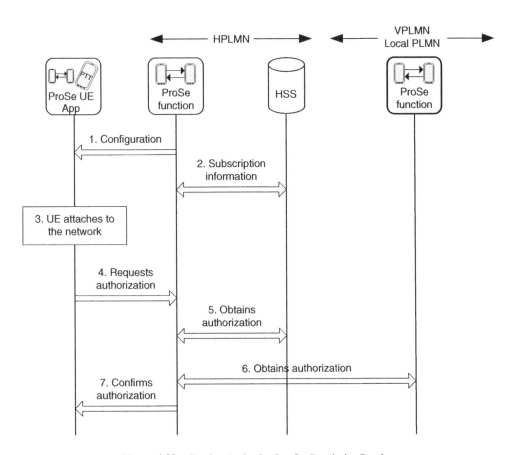

Figure 4.20   Service Authorization for Proximity Services

Final authorization of the UE is always performed by the ProSe Function in the HPLMN. The ProSe Function in the HPLMN merges authorization information from home, serving and local PLMNs. The ProSe Function in the Local PLMN, VPLMN, or HPLMN may revoke the authorization at any time. The ProSe Function in the HPLMN is notified when authorization is revoked by the Local PLMN or the VPLMN. The ProSe Function in the HPLMN notifies the UE about changes to the authorization information.

4.5   ProSe Direct Discovery

ProSe Direct Discovery is defined as the procedure used by the ProSe-enabled UE to discover other ProSe-enabled UE(s) in its proximity using E-UTRA direct radio signals using the PC5 interface without going via the network. In Release 12, ProSe Direct Discovery is supported only when the UE is served by E-UTRAN. As mentioned earlier, ProSe Direct Discovery can be a stand-alone service, (offered independent of ProSe direct communication), providing information about discovered UE(s) to certain applications running in the discoverer UE that are permitted to use this information, for example, "find a taxi nearby" or "find a coffee shop" (in these cases the discovered UE can be a device in the taxi or in the coffee shop). Additionally, depending on the information obtained, ProSe Direct Discovery can be used for subsequent actions, for example, to initiate ProSe Direct Communication.

In order to enable direct communication between UEs when they are in close proximity, UEs are required to discover each other first with the exception of Public Safety UEs. Reason for this restriction is to enforce mobile network (e.g., commercial) operator control for use of licensed spectrum for direct communication however this is not supported in Release 12. Public Safety UE(s) can communicate with each other even without ProSe Direct Discovery. In case of Public Safety UE(s), it is assumed that discovery is not necessarily needed as Public Safety personnel roughly know each other's whereabouts thus can assume another person is reachable for direct communication.

Furthermore, ProSe Direct Discovery can be used by applications as a stand-alone service (e.g., advertisement applications pushing notifications to UEs that are discovered). Such applications may not require any communication.

4.5.1   ProSe Direct Discovery Models

In case of direct discovery, following are the two possible discovery models:

1. Model A ("I am here"): In this model, the UE announces its presence to other UE(s) who are interested in reading and/or processing the messages. The UE that announces certain information (e.g., its ProSe Identity) that could be used by other UE(s) in proximity is called the "announcing UE." The UE that has the permission to discover and the interest to read and/or process messages from an "announcing UE" in proximity is called the "monitoring UE."
2. Model B ("who is there?" or "are you there?"): In this model, the UE tries to discover other UE(s) by sending a request containing certain information. The UE that transmits a request containing information like its ProSe Application ID is called as "Discoverer UE." The UE that receives and processes the request message and responds to the request is called the "Discoveree UE."

In Release 12, 3GPP specified solutions and procedures to support Direct Discovery based on "Model A" only.

4.5.2   ProSe Direct Discovery Modes

There are two types of direct discovery modes: Open and Restricted discovery. Open discovery means there is no explicit permission needed from the UE that is being discovered. Restricted discovery means explicit permission is needed from the UE that is being discovered. In Release 12, 3GPP specified solutions and procedures for "open discovery mode" only.

Use cases for open discovery may include discovering some public machine type devices (e.g., a vending machine located close to a restaurant) for which no permission is needed from the UE (within the vending machine) that is being discovered. Restricted discovery may be the most commonly used mechanism for devices used by humans to avoid privacy concerns (e.g., discovering a fireman entering a building or discovering a user walking close to a coffee shop will probably require explicit permission from the fireman or user).

4.5.3   Direct Discovery Procedure for Model A

This discovery procedure is only applied for open ProSe Discovery when the ProSe-enabled UE is served by E-UTRAN. The UE obtains service authorization from the HPLMN and serving PLMN, as described in the service authorization section.

The ProSe Application ID is used to identify applications for ProSe Direct Discovery. ProSe Application ID is globally unique and can be PLMN specific, country specific, or a global identity. Each ProSe Application ID is composed of the ProSe Application ID Name and a PLMN ID. The ProSe Application ID Name is described in its entirety by a data structure characterized by different levels, for example, broad-level business category (Level 0), business subcategory (Level 1), business name (Level 2), and shop ID (Level 3). The PLMN ID part consists of the Mobile Country Code (MCC) and Mobile Network Code (MNC) of the PLMN that assigned the ProSe Application ID. If the ProSe Application ID is country specific, then the MNC is wildcarded by "*." If it is global, then both the MCC and MNC are wildcarded by "*."

When the UE announces its presence, it initiates a "discovery request for announcing" to the ProSe Function in the HPLMN. The discovery request contains the ProSe Application ID of the application whose availability is intended to be announced. If the request is successful, it obtains the ProSe Application Code from the ProSe Function. The UE uses that for the announcing procedure over the PC5 interface.

The ProSe Application Code is a temporary code that corresponds to the ProSe Application ID and it is UE specific. It is used to enable "monitoring UE" discover the presence of the "announcing UE". Each ProSe Application Code is composed from a temporary identity and a PLMN ID that corresponds to the PLMN that assigned the ProSe Application Code. The temporary identity internal structure also follows the structure associated with the ProSe Application ID (i.e., it contains separate identifiers for each level of the Application ID) to enable partial matching at the monitoring UE side using a ProSe Application Mask or a Discovery Filter. The ProSe Application Code has a validity time that determines how long it is valid and can be used.

A ProSe Application Mask consists of one or more applicable parts of temporary identities of ProSe Application Codes to allow partial matching of ProSe Application Codes. A ProSe Application Mask is contained in a Discovery Filter.

When the UE is triggered by an application to monitor for other UEs that are in proximity and it has the authorization to monitor, it initiates a discovery request for monitoring towards the ProSe Function. The monitoring UE has to know whether it needs to send a monitoring request for PLMN specific, countrywide or a global ProSe Application ID as its scope is encoded in the Application ID.

If the request is successful, it obtains the Discovery Filter that consists of ProSe Application Code(s) and/or ProSe Application Mask(s). Then it starts monitoring for these ProSe Application Code(s) on the PC5 interface. When the UE detects that one or more advertised ProSe Application Code(s) match the given filter (i.e., are part of the filter), it reports these ProSe Application Code(s) to the ProSe Function.

In order for a ProSe Application Code to match the given filter, it requires matching of all components of the ProSe Application Code. It is considered a full match when both PLMN ID and temporary identity match with the corresponding content of the Discovery Filter. Thus, a Discovery Filter with ProSe Application Mask set to all 1's can be used for identification of full match of the service indicated in ProSe Application Code. Discovery Filter with ProSe

Application Mask that is set to 1's for the parts that should match, set to 0's for the parts that can be masked allows partial matching of as many parts of the ProSe Application Code that are contained in the ProSe Application Mask. A partial match occurs when the PLMN ID matches fully and the temporary identity matches partially with the corresponding content of the Discovery Filter. Refer to Section 4.8.6 for an illustration on how match procedure is performed using Application mask, Application Code, and a Discovery Filter.

Subsequent sections explain how direct discovery (announce, monitor, match procedures) work for non-roaming and roaming scenarios. In addition, it provides some insights about radio aspects with a focus on radio resource allocation for direct discovery procedure.

4.5.4   Radio Aspects and Physical Layer Design

The UE can participate in announcing and monitoring of discovery information in both RRC_IDLE and RRC_CONNECTED state. The UE announces and monitors its discovery information subject to the half-duplex constraint. In a half-duplex operation, transmitter and receiver cannot operate simultaneously. In the transmitter mode, UE can initiate announce procedure and in the receiver mode, it can perform monitoring. In simple terms, half-duplex implies that one party talks while the other party listens (i.e., walkie-talkie style).

Working assumption in 3GPP RAN WG is that the "announcing UE" is transmitting certain information that helps the "monitoring UE" to do a direct discovery using the new physical and transport channels that will be defined in 3GPP (refer 3GPP TR 36.843 [14]). The transport and logical channels used for direct discovery are different from those used for direct communication. In order for the UE to perform direct discovery, E-UTRA radio resources are necessary. The E-UTRA network (eNodeB) controls the allocation of resources, that is, either configures resource pools or schedules the actual resources to be used. The actual discovery messages and information are transmitted directly from the announcing UE to the monitoring UE(s) using the new physical transport channel. The announcing UE(s) selects the resources that are part of the "resource pool" for the announcement. The monitoring UE(s) have to know which channel and what time slot to look for potential transmissions of discovery messages from announcing UE(s). Monitoring UE(s) obtain this information from the network. It is assumed that this information could be read by the UE(s) from a System Information Broadcast (SIB) provided by the network.

Announcing and monitoring UE maintain the current Coordinated Universal Time (UTC). The announcing UE transmits the discovery message, which is generated by the ProSe protocol taking into account the UTC time upon transmission of the discovery message. In the monitoring UE the ProSe protocol provides the message to be verified together with the UTC time upon reception of the message to the ProSe function (for the use of UTC time see also Section 4.8.3).

4.5.5   Radio Resource Allocation for Direct Discovery

Discovery transmission resource configuration data consist of discovery period, number of subframes within a discovery period that can be used for transmission of discovery signals, and physical resource blocks (PRBs). The actual number of PRBs is yet to be decided in 3GPP. As mentioned earlier, the range for the discovery depends on the signal and radio conditions such as interference. The "maximum power transmission" level for the discovery signal to perform the discovery procedure will need to be authorized by the eNB.

There are two types of resource allocation for discovery information announcement [15].

- Type 1: A resource allocation procedure where resources for announcing of discovery information are allocated by the network that is not specific to a certain UE. In this case, the eNB provides the UE(s) with resource pool configuration that can be used for announcing discovery information. The eNB can signal this information in SIB. UE can select the radio resource from the resource pool and use this for announcing discovery information. Resource pool that is necessary for monitoring procedure can also be configured in the UE(s).
- Type 2: A resource allocation procedure where resources for announcing discovery information are allocated by the network that is specific to a certain UE. When the UE is in RRC connected mode, UE can request resource(s) for announcing discovery information and eNB can allocate resources using dedicated RRC signaling messages. The NAS layer (see 3GPP TS 24.301 [16]) within the UE will initiate the trigger for a RRC establishment procedure to obtain these radio resources by sending a Service Request message (described in Section 1.6.9). The resource pool that is necessary for monitoring procedure can also be configured in the UE(s).

When the UE is in RRC_IDLE mode, the eNB can provide Type 1 resource pool for discovery announcement in SIB that the UE can use for announcing discovery information. UE(s) that are authorized to perform discovery can use these resources for announcing discovery information. Alternatively, eNB can also indicate that it supports the D2D discovery feature but not provide any resource for discovery information announcement. This implies that the UE has to enter the RRC connected state in order to request D2D resources for discovery information announcement.

When the UE is in RRC_CONNECTED state, the UE that is authorized to perform ProSe direct discovery can request the eNB for resources. First, eNB validates whether the UE is authorized for ProSe direct discovery procedure based on the UE context information received from the MME. Second, if the UE is authorized to perform discovery, the eNB allocates necessary resources either by configuring the UE to use Type 1 resource pool or by allocating necessary resources using dedicated RRC signaling procedures.

UEs in RRC_IDLE and RRC_CONNECTED states monitor both Type 1 and Type 2 discovery resource pools as authorized. The eNB provides the resource pool configuration used for discovery information monitoring in SIB. The SIB may also contain discovery resources used for announcing in neighbor cells.

Discovery resources can be overlapping or nonoverlapping across cells. The serving cell may provide which neighbor frequencies support ProSe discovery in SIB. For a synchronized, full-overlapping, intra-frequency deployment, the eNB provides just one resource pool.

Resources allocated by the eNB may remain valid until the eNB revokes the resources using explicit RRC signaling procedures or the UE moves to idle mode.

4.5.6   Inter-frequency ProSe Discovery

In Release 12 inter-frequency ProSe Discovery (intra-PLMN and inter-PLMN discovery) will be supported for the monitoring UE. The announcing UE shall transmit discovery signals only on the serving cell carrier frequency provided it is authorized by the network. Reason

for this is that for the discovery procedure, the monitoring UE stays in the cell on whatever carrier it is camping in and monitors the resources from other carriers where announcing UEs may be present and transmitting discovery information. The monitoring UE obtains
inter-frequency resources for the monitoring procedure by reading the SIB (i.e., SIB18) from
those inter-frequency cells. In the serving cell of the monitoring UE, only the frequencies
where discovery information may come from is provided (in a SIB possibly called SIB19).

An eNodeB may broadcast in SIB a list of carrier frequencies in which the UE interested in
monitoring may aim to receive ProSe discovery signals. The serving cell for the monitoring
UE does not provide detailed ProSe resource configuration of the other carriers for the UE.
The UE may obtain resources from the ProSe carrier cell while camping on the non-ProSe
carrier cell by reading the SIB in the ProSe carrier cell. The UE can do this using a second
receiver chain, if it is a dual Rx UE.

4.5.7   Announce Procedure (non-roaming)

Example scenario: Fireman Bob and Fireman John subscribed to operator A. Bob has arrived
at a football stadium and his device is now registered in a network operated by operator A. Bob
knows that other firemen (John and others) will arrive at the football stadium. Bob would like

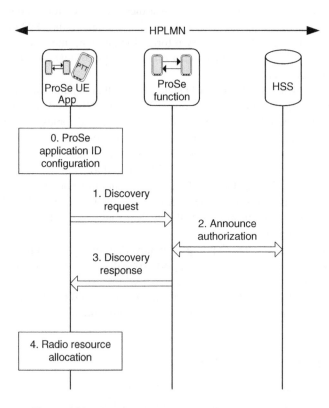

Figure 4.21   Announce request procedure (non-roaming)

to announce to the other firemen in the stadium that he has arrived, that is, Bob announces "I am here." In order for Bob to do this, his device needs to initiate the Announce procedure described in this section (Figure 4.21).

4.5.7.1   Procedure Description

Precondition
The UE is configured with an authorization policy and this indicates that the UE can perform discovery in the HPLMN. UE is also configured with a ProSe application ID that corresponds to the HPLMN.

1. On the basis of the authorization policy, UE establishes a secure connection with the ProSe Function in the HPLMN and initiates a discovery request. It indicates Announce command, ProSe Application ID, UE ID, application identity within the discovery request to initiate the Announce procedure. The application identity uniquely identifies the application itself. It should be noted that all common mobile operating systems have namespaces that identify the applications within the particular operating system. ProSe Application ID indicates the application that the UE is interested to announce. The UE Identity (e.g., IMSI) identifies the subscription data in the HSS. The UE always sends this request to the ProSe Function in the HPLMN.
2. The ProSe Function checks whether it has the UE context available. If there is no associated UE context, then it checks with HSS for authorization of discovery request against subscription data. On the basis of UE's subscription, HSS authorizes the UE and responds to the ProSe Function with subscription and authorization information. HSS provides this information to the ProSe Function. The ProSe Function creates a new UE context and determines how long the UE is authorized to perform discovery (i.e., determines the validity timer). ProSe Function retains the UE context for the duration of the validity timer. If the UE does not perform a new announce request within the duration of the validity timer, then the ProSe Function will remove the entry related to the requested ProSe Application ID from the UE context.
3. The ProSe Function responds to the UE with Discovery Response to initiate the Announce procedure. ProSe Function provides the application code and validity timer to the UE. The ProSe Application Code corresponds to the ProSe Application ID and the validity timer indicates how long the ProSe Application Code is valid in the current PLMN. The UE can perform discovery for the duration of validity timer without requesting for a new ProSe Application Code. If the UE moves to a new PLMN or the validity timer expires, then the UE needs to request a new ProSe Application Code.
4. The UE can start announcing the ProSe Application Code in the HPLMN using the radio resources authorized and/or configured by E-UTRAN for ProSe Discovery.

4.5.8   Announce Procedure (roaming)

Example scenario: Fireman Bob is subscribed to operator A. Bob has arrived at a football stadium and his device is now registered in the network operated by operator B. Bob knows other firemen (John and others) will arrive at the football stadium. Bob would like to announce

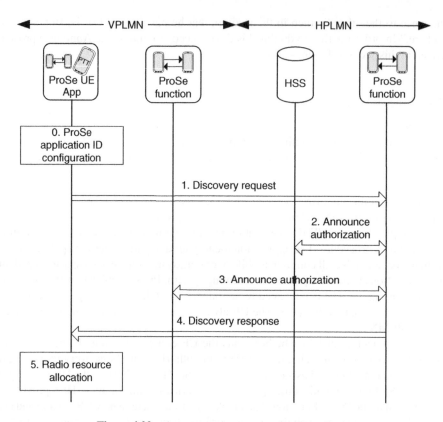

Figure 4.22   Announce request procedure (roaming)

his arrival to the other firemen in the stadium, that is, Bob announces "I am here." In order for Bob to do this, his device needs to initiate the Announce procedure described in this section (Figure 4.22).

4.5.8.1   Procedure Description

The Announce request procedure in roaming scenarios is similar to the procedure for non-roaming scenarios (described in the previous section) with the following additions:

1. The UE establishes a secure connection with the ProSe Function in the HPLMN and sends the discovery request to this ProSe Function.
2. If the UE is authorized for discovery request, the ProSe Function in the HPLMN informs the ProSe Function in the VPLMN that the UE is authorized for Announce. The ProSe Function in the HPLMN could determine the ProSe Function in the VPLMN based on configuration (e.g., based on roaming agreements with certain number of PLMNs). It also provides the ProSe Application ID, ProSe Application Code, and UE Identity to the ProSe Function in the VPLMN. The ProSe Application ID corresponds to the request from the UE, whereas the ProSe Application Code indicates the assigned code for this request. The

request also includes the UE identity information, such as International Mobile Subscriber Identity (IMSI) or Mobile Subscriber ISDN Number (MSISDN) in order to allow the ProSe Function in VPLMN to perform charging.

3. The UE may also need to be authorized by the VPLMN. In such cases, the ProSe Function in the HPLMN should obtain authorization from the ProSe Function in the VPLMN and respond to the UE.

4.5.9   Monitor Procedure (non-roaming)

Example scenario: Fireman John is subscribed to operator A. John has arrived at a football stadium and his device is now registered in a network operated by operator A. John would like to monitor the arrival of other firemen (e.g., Bob and others) in the stadium. John can also monitor firemen who are subscribed to other operator's networks (e.g., firemen subscribed to operator B or operator C), if John's device is authorized to monitor in these other networks. In order for John to monitor other devices, his device needs to initiate the monitor procedure described in this section (Figure 4.23).

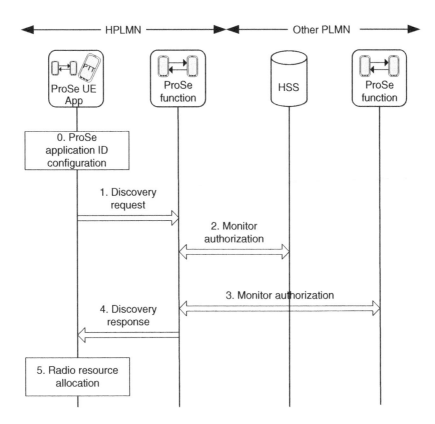

Figure 4.23   Monitor request procedure (non-roaming)

4.5.9.1   Procedure Description

Precondition

The UE is configured with an authorization policy indicating that the UE can perform discovery in the HPLMN. The UE is also configured with a ProSe Application ID that corresponds to the HPLMN.

1. On the basis of the configured authorization policy, the UE establishes a secure connection with the ProSe Function in the HPLMN and initiates a discovery request. It indicates Monitor command and includes ProSe Application ID, UE ID, application identity within the discovery request. The ProSe Application ID indicates the application the UE is interested in. If the UE is interested to monitor UE(s) in other PLMNs, that is, local PLMNs, then the UE can indicate this within the ProSe Application ID. The UE always sends the discovery request to the ProSe Function in the HPLMN.
2. The ProSe Function checks whether it has a UE context available. If there is no associated UE context, then it checks with HSS for authorization of discovery request. On the basis of UE's subscription, HSS authorizes the UE and responds to the ProSe Function with subscription and authorization information. It also contains information on the PLMNs where the UE is allowed to perform monitoring. The ProSe Function creates a new UE context and determines how long the UE is authorized to perform discovery. The ProSe Function retains the UE contexts for the duration of its validity timer.
3. If the Discovery Request is authorized and the ProSe Application ID sent by the UE in step 1 indicates other Local PLMNs, then the ProSe Function in the HPLMN contacts these Local PLMNs in order to obtain mask(s) corresponding to the ProSe Application ID Name(s). The request should also include the UE identity information (e.g., IMSI) in order to allow the ProSe Function in the Local PLMN to perform charging. If the ProSe application ID sent by the UE in step 1 indicates that it was assigned by ProSe Function in the HPLMN, then the ProSe Function in the HPLMN can directly respond to the UE.
4. If the ProSe Function of the Local PLMN stores valid ProSe Application Code(s) corresponding to the requested ProSe Application ID Name(s), then the ProSe Function of the Local PLMN returns the related mask(s) and the corresponding Time-To-Live (TTL) for each of them. If the ProSe Function of the Local PLMN does not return any mask, then it indicates that the UE is not authorized to monitor in this Local PLMN.
5. The ProSe Function in the HPLMN responds to the UE with Discovery Response including the Discovery Filter(s) and filter ID. Each Discovery Filter consists of a ProSe Application Code, one or more ProSe Application Mask(s), and a TTL timer for each Discovery Filter.
6. The UE can start monitoring using the Discovery Filter(s) in the radio resources that are authorized and configured by the PLMN(s).

4.5.10   Monitor Procedure (roaming)

Example scenario: Fireman John subscribed to operator A. John has arrived at a football stadium and his device is now registered in a network operated by operator B. John would like to monitor the arrival of other firemen (e.g., Bob and others) in the stadium. John can also

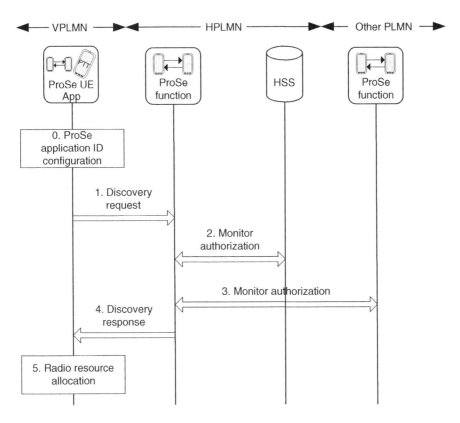

Figure 4.24   Monitor request procedure (roaming)

monitor firemen who are subscribed to other operator's networks (e.g., firemen subscribed to operator C) where John's device is authorized to monitor. In order for John to do this, his device need to initiate the monitor procedure described in this section (Figure 4.24).

4.5.10.1   Procedure Description

The Monitor request procedure in roaming scenarios is similar to the procedure for non-roaming scenario (described in the previous section) with the following additions:

1. The UE establishes a secure connection with the ProSe Function in the HPLMN and sends the discovery request.
2. If the UE is authorized to initiate a discovery request, HSS provides also the PLMN ID where the UE is registered (i.e., VPLMN ID).
3. The UE may also need to be authorized by the Local PLMNs and VPLMN. In such cases, the ProSe Function in the HPLMN should obtain authorization from the ProSe Function in the Local PLMNs and VPLMN, and respond to the UE.

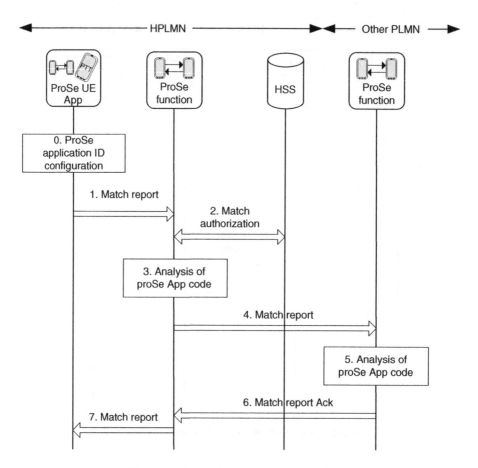

Figure 4.25   Match procedure (non-roaming)

4.5.11   Match Procedure (non-roaming)

Example scenario: Firemen John and Bob are subscribed to operator A. John has arrived at a football stadium and his device is now registered in a network operated by operator A. John would like to monitor the arrival of other firemen (e.g., Bob and others) in the stadium. John can also monitor firemen who are subscribed to other operator's networks (e.g., firemen subscribed to operator B and operator C) where John's device is authorized for monitoring. John's device has detected the arrival of Bob (Bob's device announced "I am here") by monitoring the arrival/presence of Bob's device which performed the Announce procedure. However, it is not confirmed by John's device that the Announce procedure was performed by Bob's device. In this case, John's device needs to execute the match procedure described in this section (Figure 4.25) to confirm.

4.5.11.1   Procedure Description

Precondition
John's UE initiates a monitor procedure as described in Section 4.5.9. Bob's UE performs the Announce procedure as described in Section 4.5.7.

1. John's UE detects an application code that matches the Discovery Filter but it does not have the application ID(s) already stored. In this case, John's UE establishes a secure connection with ProSe Function in the HPLMN and initiates a Match report request. This report includes ProSe Application Code(s), Discovery filter(s), and UE ID. The ProSe Application Code is the code that corresponds to the Discovery Filter that John's UE matched. John's UE always sends the match report request to the ProSe Function in the HPLMN.
2. The ProSe Function checks the context for John's UE. The authorization information also contains the PLMN for which the UE is allowed to perform discovery.
3. The ProSe Function in the HPLMN analyses the application code received from John's UE.
4. If the application code was assigned by a ProSe Function in another PLMN (i.e., Local PLMN), then the ProSe Function in the HPLMN sends a Match report request with ProSe Application Code and UE ID to the ProSe Function of the corresponding PLMN (i.e., the ProSe Function of the HPLMN of the "announcing UE," which is Bob's UE). If the ProSe Application Code was assigned by the ProSe Function in the HPLMN, then the ProSe Function in the HPLMN can directly respond to the UE. The UE identity is included to allow the ProSe Function in the Local PLMN to perform charging.
5. The ProSe Function in the Local PLMN(s) analyses the ProSe Application Code.
6. If the ProSe Application Code is confirmed, then the ProSe Function in the Local PLMN sends Match Report Acknowledgement and includes the ProSe Application ID Name(s) and validity timer(s) to the ProSe Function in the HPLMN. This message may also contain certain metadata corresponding to the ProSe Application ID Name, for example, postal address, phone number, URL, and so on.
7. The ProSe Function in the HPLMN responds to John's UE with a Match Report Acknowledgment and includes the ProSe Application ID(s) and validity timer(s). Again, this message may contain certain metadata corresponding to the ProSe Application ID Name. The validity timer(s) indicate for how long the ProSe Application ID(s) provided are going to be valid. The UE stores the mapping of ProSe Application Code(s) and corresponding ProSe Application ID(s) for the duration of their validity timer.

4.5.12   Match Procedure (roaming)

Example scenario: Firemen John and Bob are subscribed to operator A. John has arrived at a football stadium and his device is registered in a network operated by operator B. John would like to monitor arrival of other firemen (for example, Bob and others) in the stadium. John can also monitor firemen who are subscribed to other operator's networks (e.g., firemen subscribed to operator C) where John's device is authorized for monitoring. John's device has detected arrival of Bob (Bob's device announced "I am here") by monitoring the arrival/presence of

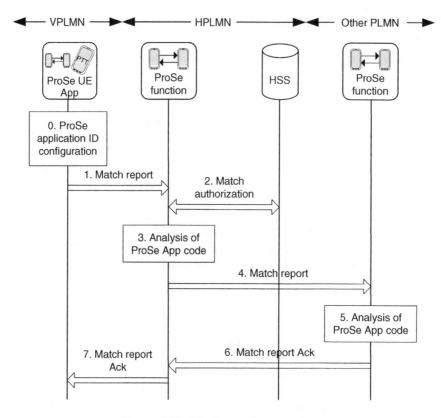

Figure 4.26 Match procedure (roaming)

Bob's device, which performed the Announce procedure. However, it is not confirmed by John's device that the Announce procedure was performed by Bob's device. In this case, John's device needs to execute the match procedure described in this section (Figure 4.26) to confirm.

4.5.12.1 Procedure Description

The Match procedure in roaming scenarios is similar to the procedure for non-roaming scenarios (described in the previous section) with the following additions:

1. The UE establishes a secure connection with the ProSe Function in the HPLMN and sends the Match report to the ProSe Function in the HPLMN. It includes the PLMN ID for the registered PLMN (i.e., VPLMN ID) and PLMN ID for the PLMN(s) where it performed monitoring (i.e., monitored PLMN ID). Monitored PLMN is the registered PLMN of the "announcing UE."
2. The UE may also need to be authorized by the Local PLMNs and VPLMN. In such cases, the ProSe Function in the HPLMN should obtain authorization from the ProSe Function in the Local PLMNs and VPLMN and respond to the UE.

4.5.13   Direct Discovery Procedure for Model B

"Model B" allows the UE to discover other UE(s) by sending a request containing certain information. The UE that transmits such a request containing certain information such as its ProSe Application ID is called the "Discoverer UE." The UE that receives, processes the request, and responds to the request is called the "Discoveree UE."

However, this procedure has been left out of scope for 3GPP Release 12, thus no solution has been specified.

4.6   ProSe Direct Communication

Direct communication or ProSe Direct Communication means communication between two or more devices (UEs) in proximity via LTE radio technology bypassing the mobile network. As mentioned earlier, discovery is not a prerequisite for direct communication between ProSe-enabled UE(s). Furthermore, direct communication is specified only for ProSe-enabled Public Safety UE(s) and it cannot be used by ProSe-enabled non-Public Safety UE(s) in 3GPP Release 12.

The two communication modes that have been considered for ProSe are network-independent direct communication and network-dependent direct communication.

Network-independent direct communication does not require any network assistance to authorize the connection, and communication is performed by using only functionality and information local to the UE(s). This mode is applicable:

- Only to preauthorized ProSe-enabled Public Safety UEs
- Regardless of whether the UEs are served by E-UTRAN or not
- For both one-to-one ProSe Direct Communication and one-to-many ProSe Direct Communication.

Network-dependent direct communication always requires network assistance to authorize the connection. This mode of operation applies:

- To one-to-one ProSe Direct Communication
- When both UEs are "served by E-UTRAN"
- For Public Safety UEs it may apply when only one UE is served by E-UTRAN.

3GPP specified a solution only for "one-to-many ProSe Direct Communication" scenarios in Release 12. This feature was prioritized in Release 12 as it was considered important for Public Safety communication.

One-to-many ProSe Direct Communication (Figure 4.27) is connectionless and it does not require control signaling over PC5. Authorization for group communication and other necessary connection parameters such as ProSe Group IP multicast addresses, ProSe Group IDs, group security data, and radio-related parameters are all configured in the UE by the ProSe Function to enable one-to-many communication. Members of the group involved in the communication share a secret from which a group secret key can be derived to encrypt user data. There is no support for specific QoS except for priority handling at the UE. Priority handling refers to the prioritization that can be done by the UE itself for D2D communication. The UE can prioritize logical channels at the MAC sublayer, referred to

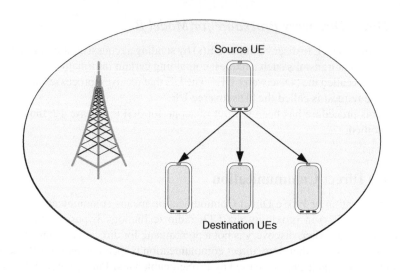

Figure 4.27   One-to-Many ProSe Direct Communication

as "logical channel prioritization" in 3GPP TS 36.321 [17]. In addition, the UE could also support dynamic scheduling, if it acts as a scheduler for direct communication but this is an implementation-specific behavior. Support for dynamic scheduling implies that the UE can prioritize the reception based on the source UE, for example, UE A is receiving traffic from UE B, UE C, UE D; UE A can prioritize UE B over UE C and UE D.

4.6.1   Radio Aspects and Physical Layer Design

Based on the status of discussions in 3GPP at the time of writing this book it is likely that physical and transport channels used for direct communication are different from that used for direct discovery. A new logical channel and a new transport channel are going to be defined for proximity services (refer to 3GPP TR 36.843 [14] for a discussion of these new channel types). The newly defined logical channel is called ProSe Communication Traffic Channel (PTCH) and the new transport channel is called ProSe Communication Shared Channel (PSCH).

4.6.1.1   ProSe Communication Traffic Channel (PTCH)

A PTCH is a point-to-multipoint channel, for transfer of user information from one UE to other UE(s). A PTCH can exist in both transmission and reception side. This channel is used only by Prose Direct Communication-capable UEs.

4.6.1.2   ProSe Communication Shared Channel (PSCH)

The PSCH is unidirectional. Thus, there are separate PSCH channels in "transmit" and "receive" directions. Characteristics of the PSCH are as follows:

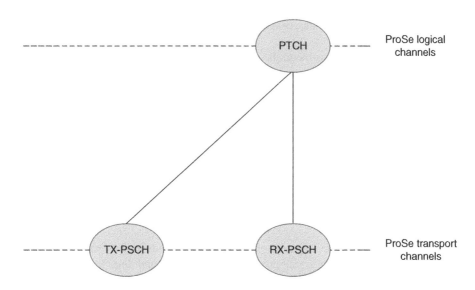

Figure 4.28   Mapping between ProSe logical transport channels

- Support for broadcast transmission
- Support of dynamic resource allocation for mode 1 resource allocation (mode 1 is described later)
- Ability to handle collision risk due to support of UE autonomous resource selection for mode 2 resource allocation (mode 2 is described below)
- No support for Hybrid Adaptive Repeat and Request (HARQ) feedback (see Section 4.12 Terms and Definitions at the end of this chapter for additional information related to HARQ).

The PTCH logical channel maps to transmit and receive PSCH transport channels for ProSe Direct Communication (see Figure 4.28).

For ProSe transmission, PTCH is mapped to Tx-PSCH. For ProSe reception, PTCH is mapped to Rx-PSCH.

4.6.2   Radio Resource Allocation for Direct Communication

UE(s) in-coverage and out-of-coverage should be aware of the radio resource (time and frequency) for direct communication. The transmitter UE transmits the so-called Scheduling Assignment (SA) to indicate the resources it is going to use for data transmission to the receiver UEs.

There are two modes of operation for ProSe-enabled UE(s) [15] – Mode 1 and Mode 2. The eNB can configure in which mode of operation the UE operates when it is in-coverage. When the UE is in-coverage, UE must use the mode as configured by the eNB unless an exceptional situation occurs. When the UE is out-of-coverage, UE can only use mode 2. The UE is considered to be in-coverage, if it is in RRC_CONNECTED state or is camping on an E-UTRA cell in RRC_IDLE state.

Mode 1 (eNB Scheduled Resource Allocation)

When the UE operates in this mode the eNB schedules resource allocation. The UE needs to be in RRC_CONNECTED state in order to communicate. The UE requests the eNB for transmission resources. The eNB schedules transmission resources for transmission of SA(s) and actual data. The UE sends a scheduling request (D-SR or random access) to the eNB followed by a Buffer Status Reports (BSRs). Based on the BSR, the eNB can determine that the UE has data for ProSe Direct Communication transmission and estimates the resources needed for transmission. The eNB schedules the specific resource(s) to use for SA transmission. The specific resource assigned by the eNB is within the resource pool provided to the UE.

Mode 2 (UE Autonomous Resource Selection)

This is a mode in which the UE autonomously selects resources for communication. The UE selects resources from resource pools to transmit a SA and data. In principle, the UE can engage in direct communication when it is in RRC_IDLE state, if the eNB configures the UE to conduct communication in Mode 2. This is logical because the UE is not required to establish RRC connection with the network in order to engage in communication when it is configured for Mode 2.

When the UE is out-of-coverage, the resource pool used for transmission and reception of SA is preconfigured in the UE. When the UE is in-coverage, resource pools used for transmission and reception of SA can be configured by the eNB using RRC dedicated signaling or using broadcast signaling. If the UE is in-coverage, it should use only the mode indicated by eNB configuration unless one of the exceptional cases occurs (e.g., a radio link (re-)establishment failure).

All UEs (Mode 1 and Mode 2 UEs) are provided with a resource pool (time and frequency) in which they attempt to receive SAs.

When the UE is in RRC_IDLE state, the eNB can provide a Mode 2 transmission resource pool in SIB. UEs that are authorized for Prose Direct Communication use these resources for ProSe Direct Communication in RRC_IDLE. The eNB can also indicate in SIB that it supports ProSe Direct Communication but does not provide resources for it. UEs need to enter RRC_CONNECTED state to perform ProSe Direct Communication transmission.

When the UE is in RRC_CONNECTED state and authorized for ProSe Direct Communication transmission, the UE indicates to the eNB that it wants to perform ProSe Direct Communication transmission. The eNB validates whether the UE is authorized for ProSe Direct Communication transmission using the UE context received from MME. Then, the eNB configures the UE in RRC_CONNECTED by dedicated signaling with a Mode 2 resource allocation transmission resource pool that may be used without constraints while the UE is RRC_CONNECTED. Alternatively, the eNB may configure a UE in RRC_CONNECTED by dedicated signaling with a Mode 2 resource allocation transmission resource pool that the UE is allowed to use only in exceptional cases (e.g., in case of radio link (re-)establishment failure). For normal cases, the UE should rely on Mode 1 resource allocation.

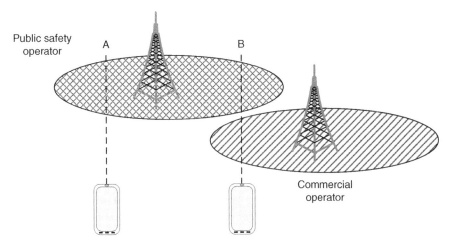

Figure 4.29   Inter-frequency ProSe communication

4.6.3   Inter-frequency ProSe Communication

Figure 4.29 depicts a scenario where a Public Safety operator and a commercial operator cooperate to provide Public Safety services. Both the Public Safety operator as well as the commercial operator have their own spectrum and deploy their own base stations. However, Public Safety UE(s) are allowed to camp on the commercial operator's cells in areas where the Public Safety operator does not offer LTE coverage. This could be realized by Equivalent Home PLMNs (EHPLMN list) configured in the UE or by secondary PLMN IDs broadcast by the commercial operator's cells. Similar cases can occur in commercial roaming scenarios (e.g., the Public Safety operator is the HPLMN and the commercial operator is the VPLMN) [18, 19].

In Release 12 [20], it can be assumed that all ProSe Communication is performed on a single carrier (the ProSe carrier), which is known to the UE by configuration. A UE that is camping on a non-ProSe carrier but interested in ProSe Communication on the ProSe carrier should attempt to find/detect cells on the ProSe carrier and thereby determine whether it is "in-coverage."

If the UE is RRC-CONNECTED, it sends a ProSe indication to its serving eNB when it wants to perform ProSe Communication. The indication contains the intended ProSe frequency. The eNodeB may configure an inter-frequency RRM measurement on the ProSe carrier and based on the measurement report, it triggers inter-frequency mobility to that ProSe carrier once the UE enters a cell using the ProSe carrier. If the UE is RRC IDLE, it may reselect to the ProSe carrier once it detects a suitable cell. If the UE detects a suitable cell on the ProSe carrier, the UE should no longer use the resources configured in the Universal Integrated Circuit Card (UICC) but instead use the resources configured by eNB providing coverage for the ProSe carrier. If the eNB does not provide the UE with ProSe resources in SIB or dedicated signaling, the UE should stop any ProSe operation in order not to harm the network and existing connections. In Release 12 the eNB providing coverage in the non-ProSe carrier cannot configure resources belonging to the ProSe carrier. In other words, the commercial operator's eNB cannot configure resources controlled by the Public Safety operator's eNB. The ability for one eNB to configure resources controlled by the other eNB is called cross-carrier scheduling of resources. Cross-carrier scheduling of resources will not be supported in Release 12. The UE performs mobility toward the ProSe carrier first and the eNB providing coverage in the ProSe carrier configures resources after the mobility event.

4.6.4   IP Address Allocation

The UE can support either IPv4 or IPv6 addresses. Following are the possible options in order to support one-to-many ProSe Direct Communication:

– The UE can be configured to use IPv6 on the direct link. In this case, the UE autoconfigures a link local IPv6 address following procedures defined in IETF RFC 4862 [21]. This address can only be used as the source IP address for one-to-many ProSe Direct Communication.
– The UE can be configured to use IPv4 for a certain Group one-to-many ProSe Direct Communication. In this case, it uses the configured IPv4 address for the Group one-to-many communication. If it is not configured with an address for the Group, it uses dynamic configuration of IPv4 link local addresses according to IETF RFC 3927 [22] to obtain an IPv4 address.

4.6.5   One-to-Many Communication (Transmission)

Example scenario: Firemen Bob and John have discovered each other (e.g., using the announce, monitor, and match report procedure). Now, Bob is trying to communicate with John and other firemen. Bob's device has to perform the procedure described in this section to

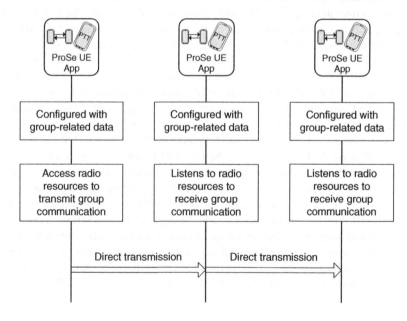

Figure 4.30   One-to-many ProSe Direct Communication transmission

initiate transmission of data (see Figure 4.30). If Bob and John discover each other by other means (e.g., they are able to see each other), they can also try to communicate without their devices executing the Direct Discovery procedure.

4.6.5.1   Precondition

The UE is configured with the necessary authorization information and radio resource-related information for one-to-many ProSe Direct Communication.

4.6.5.2   Procedure Description

1. The UE obtains the necessary group context (ProSe Layer 2 Group ID, ProSe Group IP multicast address) to transmit IP packets and also the radio resource related parameters used for Direct Communication.
2. The "transmitting UE" finds the appropriate radio resource to conduct one-to-many ProSe Direct Communication. This depends on whether the UE is in-coverage or out-of-coverage and the mode of operation.

   The PDU passed for transmission to the AS is associated with a Layer 3 PDU type. The supported Layer 3 protocol data types are (i.e., these types of IP packets area allowed to be transmitted): IP and Address Resolution Protocol (ARP) (see RFC 826 [23]). The packet is associated with the corresponding Source Layer 2 ID and Destination Layer 2 ID. The Source Layer 2 ID is set to the ProSe UE ID assigned by the ProSe Key Management Function. The Destination Layer 2 ID is set to the ProSe Layer 2 Group ID.
3. The originating UE sends the IP packet to the IP multicast address using the ProSe Layer 2 Group ID as Destination Layer 2 ID.

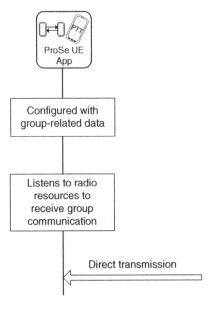

Figure 4.31   One-to-many Direct Communication reception

4.6.6   One-to-Many Communication (Reception)

Example scenario: Firemen Bob and John have discovered each other. Now, Bob is trying to communicate with John and other firemen. Bob's device has to perform the procedure described in this section in order to listen to John and other firemen (Figure 4.31). If Bob and John discover each other by other means (e.g., they are able to see each other), they can also try to communicate without their devices executing the Direct Discovery procedure.

4.6.6.1   Precondition

The UE is configured with the necessary authorization information and radio resource-related information for one-to-many ProSe Direct Communication.

4.6.6.2   Procedure Description

1. The UE obtains the necessary group context (ProSe Layer 2 Group ID, ProSe Group IP multicast address) to transmit IP packets and also the radio resource-related parameters used for the Direct Communication.
2. The "receiving UE" uses the allocated radio resource to receive one-to-many ProSe Direct Communication.
3. The receiving UE filters out the received frames based on the ProSe Layer 2 Group ID contained in the Destination Layer 2 ID and if it matches one of the configured Group IDs, it delivers the enclosed packet to upper layers. The IP stack filters the received packets based on the Group IP multicast address.
4. The PDU passed to the upper layers is associated with a Layer 3 PDU type. As mentioned earlier, supported Layer 3 protocol data types are: IP and ARP.

4.6.7   Direct Communication via ProSe Relay

D2D communication via a ProSe Relay can help enhance capacity and coverage in urban environments. In a dense urban environment, cellular coverage can be spotty, especially indoors, because most base stations are typically placed at or near intersections where they provide the greatest coverage to outdoor users. Often coverage can be poor in the middle of large structures, as signals may be blocked by the structure of a building. Poor indoor coverage can be solved with additional small cell deployments, but this is costly and sometimes not possible.

That is where D2D communication provides a novel approach to solve poor coverage conditions, because users with a strong signal can help nearby users with a weak signal by acting as network-to-UE relay. This greatly enhances coverage and restores service for users near the interior of buildings (concrete and steel). However, the solution for ProSe (network-to-UE) relays was not specified in 3GPP Release 12.

4.7   EPC-Level ProSe Discovery

EPC-level ProSe Discovery refers to the procedure that allows two UEs to discover each other with assistance from the network when they are in proximity. It could also be referred to as network-assisted discovery. This procedure can be used by the UEs as a stand-alone service. It can also be used in conjunction with direct communication or in conjunction with WLAN direct discovery and communication. It is applicable for both E-UTRAN direct communication and WLAN direct communication. The Wi-Fi Peer-to-Peer (P2P) specification [24] defines an architecture and set of protocols that facilitate direct discovery and communication using

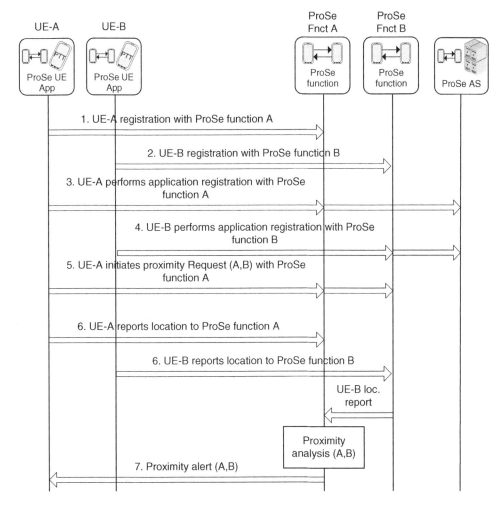

Figure 4.32   EPC-level Discovery

the IEEE 802.11 technology [25]. When EPC support for WLAN direct discovery and communication is requested as part of the EPC-level ProSe Discovery procedure, the additional parameters that are necessary to enable WLAN direct discovery and communication are provided as assistance parameters. The assistance information is designed to expedite WLAN direct discovery and communication. The content of the assistance information depends on the technology used on the WLAN direct link.

4.7.1   EPC-Level ProSe Discovery Procedure

This procedure involves the UE(s) registering themselves and the ProSe application(s) with the ProSe Function and requesting for proximity alert when the registered UE(s) are in proximity. Thus, it also involves the network determining the UE location based on OMA SUPL or other location determination technologies. When the UE(s) are in close proximity, the ProSe Function alerts the registered UE(s) accordingly.

Example scenario: Firemen Bob and John are subscribed to operator A. Bob has arrived at a football stadium and his device is now registered in a network operated by operator B. Bob knows that other firemen (John and others) will arrive at the football stadium. Bob would like to know when John and other firemen arrive. In order for Bob and John to discover each other, Bob's device and John's device should register with the ProSe Function for Proximity alert. The high-level procedure is shown in Figure 4.32 (John's device is referred to as UE-A, Bob's device is referred to as UE-B).

1. UE-A initiates the registration procedure to obtain EPC-level ProSe Discovery services. UE-A registers with the ProSe Function A residing in its home PLMN.
2. UE-B does the same.
3. If a specific application (e.g., a Public Safety application) triggered the UE to initiate EPC-level ProSe Discovery, the UE performs the application registration procedure by registering the application with the ProSe Function residing in the HPLMN. Thus, if there was such an application requesting for EPC-level ProSe Discovery, UE-A initiates registration of the corresponding application with the ProSe Function residing in its HPLMN.
4. Similarly, UE-B initiates registration of application with the ProSe Function B residing in its HPLMN.
5. UE-A makes a proximity request for UE-B, that is, requests that it be alerted for proximity with UE-B (possibly indicating a time window during which the request is valid). In response, the ProSe Function in the HPLMN of UE-A requests location updates for UE-A and UE-B. These location updates can be periodic, based on a trigger, or a combination of both. To request location updates for UE-A, the ProSe Function contacts a SLP in the HPLMN of UE-A. To request location updates for UE-B, the ProSe Function of UE-A contacts the ProSe Function of UE-B, which in turn requests location updates for UE-B.
6. Location information for UE-A and UE-B is reported to their respective ProSe Functions intermittently. ProSe Function B forwards UE-B's location updates to ProSe Function A based on the conditions set by ProSe Function A. Whenever ProSe Function A receives location updates for UE-A and/or UE-B, it performs proximity analysis on UE-A and UE-B's locations.

7. When ProSe Function A detects that the UEs are in proximity, it informs UE-A that UE-B is in proximity by sending "Proximity Alert." If the proximity request included a request for WLAN assistance information, it provides UE-A with assistance information for WLAN direct discovery and communication with UE-B. ProSe Function A also informs ProSe Function B, which in turn informs UE-B of the detected proximity. Similarly, if the proximity request included a request for WLAN assistance information, it provides UE-B with assistance information for WLAN direct discovery and communication with UE-A. To assist WLAN direct discovery and communication, the assistance information includes parameters such as:

   – SSID: The Service Set Identifier (SSID) to use for Wi-Fi P2P operation. To be compliant with the Wi-Fi P2P specification [24], the SSID should be in the form "DIRECT-ab" where a, b are two random characters.
   – WLAN Secret Key: The preshared key to be used by UEs to secure their Wi-Fi P2P communication. This is used by UEs as the Pairwise Master Key (PMK).
   – Group Owner (GO) indication: This indicates whether the UE supports GO functionality specified in the Wi-Fi P2P specification [24]. The UE implementing this functionality essentially becomes an Access Point that transmits Beacons with the P2P Information Element and accepts associations from other Wi-Fi P2P devices or from legacy Wi-Fi devices (those not implementing the Wi-Fi P2P functionality). If not set, the UE should behave as a Wi-Fi P2P client that attempts to discover and associate with a GO.
   – P2P Device Address of self: This is the WLAN Link Layer ID to be used by UE to advertise itself. A UE implementing the GO and indicates the WLAN Direct device from which the GO should accept WLAN association requests. Association requests from all other WLAN devices should be rejected by GO.
   – P2P Device Address of peers: This is the WLAN Link Layer ID to be used by UE to discover peer UEs. A UE implementing the GO should accept WLAN association requests only from devices that are in this list.
   – Operation channel: The channel on which Wi-Fi P2P discovery and communication should take place.
   – Validity time: The time period during which the content provided in the assistance information is valid.

4.7.2   User Equipment Registration

As explained in the previous section (steps 1 and 2), the UE needs to register with the ProSe Function. This section (Figure 4.33) outlines the steps for UE registration toward ProSe Function.

1. The UE registers with the ProSe Function by sending ProSe Registration Request with IMSI and WLAN Link Identifier (WLLID). The UE includes the permanent WLLID in case it supports permanent WLLID and it intends to use EPC support for WLAN direct discovery and communication. Otherwise, it obtains a temporary WLAN Link Layer ID from the ProSe Function as part of the Proximity Request procedure.
2. The ProSe Function may interact with the HSS in order to authenticate the user and check whether the user is authorized for ProSe. Alternatively, all user settings related to authentication and authorization for ProSe may be configured locally in the ProSe Function.

3. The ProSe Function generates an EPC ProSe User ID for the UE, stores the EPC ProSe User ID together with the user's IMSI, and responds to the UE by sending a UE Registration Response that includes the EPC ProSe User ID.

4.7.3   Application Registration

When a user registers with a third-party application server, the user is designated an Application Layer User ID (ALUID) (e.g., ALUID_A for user A). To activate ProSe features such as EPC-level ProSe Discovery for a specific application, the UE registers the application with the ProSe Function, as illustrated in Figure 4.34.

1. The UE sends Application Registration Request including EPC ProSe User ID, Application ID, Application Layer User ID to the ProSe Function to register an application for ProSe. The Application ID is used to identify the third-party Application Server.
2. The ProSe Function uses the EPC ProSe User ID to retrieve user's profile, checks that the requested application is on the stored list of authorized Application IDs, and sends ProSe Registration Request to the Application Server indicating that a user of this application (identified by ALUID) has requested to use ProSe services for that application. If the Application Server accepts the request, it stores the user's context information such as Application Layer User ID and EPC ProSe User ID and ProSe function ID. ProSe function ID identifies the corresponding ProSe Function.
3. The Application Server sends ProSe Registration Response to the ProSe Function indicating the status of the registration (success or failure).

The ProSe Function sends Application Registration Response including the Allowed Range to the UE. It also indicates the status of the registration. The Allowed Range parameter contains

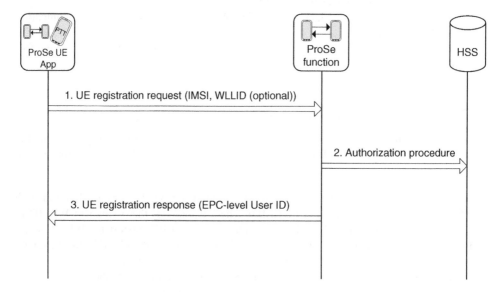

Figure 4.33   EPC-level Discovery–UE registration

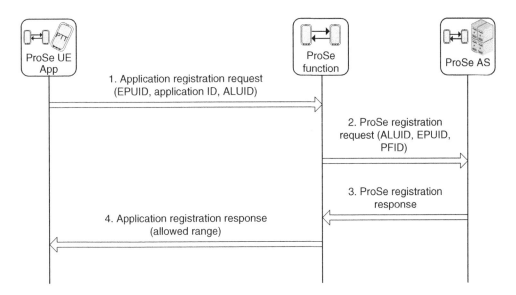

Figure 4.34   EPC-level Discovery–Application registration

the set of range classes that are allowed for this application. The range class is an integer in the 0–255 range and each integer corresponds to a certain allowed range (i.e., 1–up to 50 m, 2–up to 100 m, 3 up to 200 m, 4 up to 500 m, 5 up to 1000 m, other integers are not defined).

Further details about EPC-level discovery procedures can be found in 3GPP TS TS23.303 [3] and 3GPP TS 24.334 [5].

4.8   Other Essential Functions for Proximity Services

4.8.1   Provisioning

Provisioning parameters for ProSe Direct Discovery and ProSe Direct Communication may be configured in the USIM(see 3GPP TS 31.102 [26]), Mobile Equipment (ME) or in both. The ME provisioning parameters indicate whether ProSe Direct Discovery and ProSe Direct Communication are permitted to be used by the UE with the selected USIM. ProSe Direct Discovery and ProSe Direct Communication are accessible only when a USIM authorized for ProSe Direct Discovery and ProSe Direct Communication is selected by the user. The ME provisioning parameters shall not be erased when a USIM is deselected or replaced. If both the USIM application and the ME contain the same provisioning parameters, the USIM parameter values take precedence.

The operator configures a Public Safety ProSe-capable UE with provisioning parameters for discovery and communication, thus an explicit connection to the ProSe Function is not required in this case.

A ProSe-capable UE obtains authorization from the HPLMN to use ProSe Direct Discovery, Direct Communication, or communication via ProSe Relay on a per PLMN basis. The UE is configured with separate policies for performing ProSe Direct Discovery Monitoring and Announcing procedures. The authorization policy can indicate whether the UE is authorized to perform Announcing or Monitoring in the PLMN in which the authorization policy applies.

In addition, the UE is configured with an authorized discovery range to perform the Announcing procedure. This indicates the announcing range for ProSe Direct Discovery in the PLMN in which this announcing authorization policy applies. Detailed configuration information can be found in 3GPP TS 24.333 [27]. Announcing range is the maximum transmit power level that the UE is authorized to use for announcing for ProSe direct discovery in the PLMN in which this announcing authorization policy applies.

4.8.2   Subscription Data

The subscriber is required to obtain a subscription in order to avail proximity services described in this chapter. The subscriber can obtain subscription for individual features such as discovery, communication, or obtain subscription for all features. Thus, following features can be subscribed independently:

- Subscription for ProSe Direct Discovery
- Subscription for one-to-many ProSe Direct Communication (applicable only for ProSe-enabled PS UE(s) in 3GPP Release 12)
- Subscription for EPC-level ProSe Discovery
- Subscription for EPC-assisted WLAN direct discovery and communication
- Subscription for ProSe UE-to-Network Relay (applicable only for ProSe-enabled PS UEs).

Subscription information will be stored in the subscriber's profile in the HSS. This profile may also contain additional parameters such as list of PLMN(s) where the UE is authorized to perform one-to-many ProSe Direct Communication or to perform announce, monitor, or both.

4.8.3   Security

4.8.3.1   Security for Direct Discovery

In order to support ProSe open discovery, the specified security solution mitigates the replay and impersonation attacks. During the Announce procedure, the code announced by the UE is integrity protected. The system ensures that there is a UTC time parameter associated with each discovery slot that is known to both the announcing and monitoring UE(s). Integrity protection using Message Integrity Check (MIC) enables the ProSe Function to verify that the announcing UE was indeed authorized to announce the ProSe Application Code at that time instance.

4.8.3.2   Security for Direct Communication

Security for one-to-many ProSe Direct Communication consists of bearer level and media plane security mechanisms.

4.8.3.3   Bearer Level Security

The bearer level security uses an Identity-Based Crypto approach. The UE needs to have an algorithm identity and a ProSe Group Key (PGK) preprovisioned for each group that they belong to. From this key, the UE that wishes to broadcast encrypted data generates a ProSe Traffic Key (PTK). The parameters used in this generation ensure that PTKs are unique for each UE and this is transferred in the header of user data packets.

From the PTK, a UE derives the needed ProSe Encryption Key (PEK) to be able to encrypt the data. The UE can protect the data to be sent with the relevant keys and algorithms at the bearer level. A receiving UE would need to derive the PTK using the information in the bearer header and the PEK is used to decrypt the data.

In order to enable protection of traffic between UEs, ProSe-enabled Public Safety UEs shall implement the ciphering algorithms EEA0, 128-EEA1, and 128-EEA2. In addition, the UE may implement 128-EEA3 for ciphering one-to-many traffic. It should be noted that use of EEA0 implies that there is no encryption.

4.8.3.4 Media Plane Security

The integrity and confidentiality protection for one-to-many communications is achieved using the Secure Real Time Transport Protocol (SRTP) and Secure RTCP (SRTCP).

A Public Safety UE is provisioned by a Key Management System (KMS) with key material associated with its identity. If required, the Generic Bootstrapping Architecture (GBA) is used to bootstrap the security of the connection between the UE and the KMS. The KMS also provisions the group manager with keying material for the identity of groups which it manages.

The group manager distributes Group Master Keys (GMK) to UEs within the group. The GMK is encrypted to the user identity associated to the Public Safety UE and signed using the (key associated with the) identity of the group, an identity which the group manager is authorized to use by the KMS. Once a GMK has been distributed within the group, UEs are able to setup group communications. The initiating UE generates, encrypts, and transmits a Group Session Key (GSK) to group members. This transmission is encrypted using the GMK and may be authenticated, allowing the origin of the transmission to be verified. The distribution process may also be performed over direct communication.

MIKEY preshared key message is used for session key distribution for groups. The MIKEY message sent from one UE to another directly or via the network encapsulates the session key within the GMK. The MIKEY message may be signed using the identity of the initiating UE. This provides a network independent mechanism for the GSK distribution.

The group members share a GSK during the session setup procedure. The GSK is the Secure Real-Time Transport Protocol (SRTP) master key to provide media security of the one-to-many communication. The session may only last for a single transmission or it may be maintained for a longer period to allow many members of the group to efficiently communicate. The GSK is used as the SRTP master key to provide media security for one-to-many communication.

4.8.3.5 Security for EPC-Level Discovery

The ProSe Functions A and B of two UEs UE A and UE B trying to discover each other may belong to different PLMNs (PLMN A and PLMN B). The Application Server discloses the user identity (EPC ProSe UE ID) of a particular target UE to a ProSe Function of a PLMN A so that it can request the Prose Function of PLMN B that manages the target UE for location information.

The security risk is that it would take just one proximity request from the UE for the ProSe Function to actually keep a record of [EPC ProSe UE ID, ALUID, Application ID] mapping and have the capability to send a proximity request at a later time although the UE did not actually request for proximity alert. This could lead to massive surveillance of users by other PLMNs that may be very difficult to detect by the PLMN which serves particular ProSe UEs.

To mitigate this risk, 3GPP introduced the Application Server signed proximity request procedure. UE A does not sign the proximity request sent to its ProSe Function A, but trusts the Application Server to control the authorization of the proximity request sent on its behalf. The authorization criteria can be based on detection mechanisms, for example, very high volume of incoming proximity requests from a ProSe Function that does not match with the frequency usage of the ProSe Application by the users or it can be based on a presence detection mechanism over the PC1 interface. ProSe Function A requests an authorization to the Application Server for each proximity request. The Application Server returns parameters, specifying which operations are authorized (e.g., authorized to send only one request, authorized to send X requests until particular date, etc.).

This is to ensure that ProSe Function B is assured of the authenticity of the proximity request received from ProSe Function A by verifying the signature with a verification key from the Application Server. The token verification key is fetched over the PC2 interface between ProSe Function B and Application Server.

Details about security for ProSe Direct Discovery, ProSe Direct Communication, and EPC-level Discovery can be found in 3GPP TS 33.303 [28].

4.8.4   Charging

This section provides an overview of the agreed charging principles that are documented in the Technical Report for Proximity Services charging in 3GPP TR 32.844 [29] and the agreements that will be documented in the Technical Specification for Proximity Services Charging in 3GPP TS 32.277 [30].

Both online and offline charging are supported for ProSe Direct Discovery and EPC-level Discovery procedures. Online charging is not applicable for ProSe one-to-many Direct Communication for Public Safety use cases.

Online charging for PS may use the Immediate Event Charging (IEC) principle as specified in 3GPP TS 32.299 [31].

4.8.4.1   Charging for ProSe Direct Discovery

In order to support offline charging in case of ProSe Direct Discovery scenarios, the charging information regarding the use of ProSe services is collected by the ProSe Functions in HPLMN, VPLMN, and Local PLMNs. Inter-operator charging is supported. The charging information is collected when a UE performs ProSe Direct Discovery such as Announcing and Monitoring.

When a Charging Event is reported to the Charging Data Function (CDF), it includes details such as Subscription-id (e.g., the IMSI), PLMN ID, specific ProSe Direct Discovery Model (e.g., "Model A," "Model B"), specific ProSe UE's role used (e.g., Announcing UE,

Monitoring UE), specific ProSe functionality used (e.g., Announcing, Monitoring, Match), allocation of a ProSe Application Code to an Announcing UE and the associated period, allocation of a set of Filters for a Monitoring UE and the associated period, Match of the ProSe Application Code at a Monitoring UE and the timestamp, ProSe Application ID for ProSe Direct Discovery Announce request and monitoring. Following are the trigger conditions for charging information collection:

- ProSe Function responds to a Direct Discovery request with command (Announce or Monitor)
- ProSe Function responds to a Monitor request
- ProSe Function responds to a Announce Authorization message
- ProSe Function responds to a Match Report message.

For ProSe Direct Discovery, online charging is applicable for the Announce and Monitor procedures.

4.8.4.2   Charging for EPC-Level Discovery

In order to support offline charging in case of EPC-level Discovery scenarios, the following condition should trigger charging events and finally the generation of Charging Data Records (CDR):

- Charging data related to EPC-level Discovery Proximity request
- Charging data related to EPC-level Discovery Proximity Alert
- Charging data related to EPC-level Discovery Proximity request cancelation.

The chargeable events are defined as follows:

- Proximity request: the first charging event that triggers the creation of a new CDR and the corresponding information such as EPC ProSe User ID, Application Layer User IDs (ALUID), Application ID, time window, range, and UE's location are captured.
- Proximity Request Renewal: the next charging event that causes updates to an already open CDR for the corresponding Proximity Request with new location of the UE and time window.
- Proximity Request Reject: the last charging event that causes closure of an already open CDR for the corresponding Proximity Request. Reason for Proximity request reject is also captured.
- Proximity Request Cancelation: the last charging event that causes closure of an already open CDR for the corresponding Proximity Request. An indication whether Proximity Alert was sent is also captured.

Online charging is applicable for the Proximity Request procedure in case of ProSe EPC-level Discovery.

4.8.4.3 Charging for One-to-Many ProSe Direct Communication

In order to support offline charging for one-to-many ProSe Direct Communication, following high-level principles apply:

- In the ProSe Direct communication architecture, the Charging Trigger Function (CTF) is located in the UE and the ProSe Function. The CTF functional block located in the UE is called as the Accounting Metrics Collection (AMC). The CTF functional block located in the ProSe Function is called Accounting Data Forwarding (ADF) function (refer to Section 1.6.12 for an overview of the logical charging architecture).
- The charging operation results (e.g., errors in usage information collection or reporting) should not affect UE's use of the ProSe Direct Communication service.
- The ProSe Direct Communication usage information is stored securely in the UE when the UE is "out-of-coverage" and is uploaded securely to a location configured by the ProSe Function when the UE moves back to coverage.
- The ProSe Function is able to control the UE uploading behavior using service authorization and provisioning mechanism.
- When the UE is "in-coverage", it accesses the ProSe Function in the HPLMN.
- When the UE is "out-of-coverage", it uses preconfigured information, either from ME or UICC, or configuration received while in-coverage, for data usage logging and uploading control.
- In case of roaming, inter-PLMN charging is supported.

Online charging does not apply for ProSe one-to-many Direct Communication that is designed for Public Safety use cases. One reason is that for ProSe Direct Communication procedures, the UE does not involve the network when starting or terminating the direct communication.

In addition, there is no requirement for any credit control for the use of ProSe Direct Communication. The Public Safety regulators also have requirements that the charging control shall not prevent the UE from using the ProSe Direct Communication for Public Safety. Furthermore, the UE can communicate when it is in-coverage or out-of-coverage. Therefore, support for online charging is not deemed necessary for ProSe one-to-many Direct Communication for Public Safety use cases.

4.8.5   ProSe-Related Identifiers

4.8.5.1 ProSe UE ID

This is a link layer identifier assigned by the ProSe key management function within the ProSe Function that uniquely represents the UE in the context of one-to-many ProSe Direct Communication for a group. It is used as a source Layer 2 ID in all the packets the UE sends for ProSe Direct Communication.

4.8.5.2 ProSe Layer 2 Group ID

The ProSe Layer 2 Group ID is a link layer identifier that identifies the group in the context of one-to-many ProSe Direct Communication. It is used as a destination Layer 2 ID in all the packets the UE sends to this group for one-to-many ProSe Direct Communication.

4.8.5.3   Source Layer 2 ID

This is to identify the sender of the packet at the PC5 interface. The Source Layer 2 ID is used for identification of the receiver RLC Unacknowledged Mode (RLC UM) entity; Access Stratum signaling is not required to configure Source Layer 2 ID in the UE. This is provided by the higher layers.

4.8.5.4   Destination Layer 2 ID

This is to identify the target of the packet at the PC5 interface. The Destination Layer 2 ID is used for filtering of packets at the MAC layer. The Destination Layer 2 ID may be a broadcast, groupcast, or unicast identifier.

It should be noted that Access Stratum signaling is not required for group formation and to configure Source Layer 2 ID and Destination Layer 2 ID in the UE. This information is provided by upper layers.

4.8.5.5   SA L1 ID

This is to identify the Scheduling Assignment (SA) at the PC5 interface. SA L1 ID is used for filtering of packets at the physical layer. The SA L1 ID may be a broadcast, groupcast, or unicast identifier. In case of group- and unicast, the MAC layer will convert the higher layer ProSe ID (i.e., ProSe Layer 2 Group ID and ProSe UE ID) identifying the target (Group or UE) into two bit strings out of which one can be forwarded to the physical layer and used as SA L1 ID whereas the other is used as Destination Layer 2 ID. For broadcast purposes, the MAC layer indicates to the Physical layer that it is broadcast transmission using a predefined SA L1 ID in the same format as for groupcast and unicast.

4.8.5.6   ProSe Application ID

This is an identity used for ProSe Direct Discovery, identifying application-related information for the ProSe-enabled UE. Each ProSe Application ID is globally unique, for example, it unambiguously identifies a service across all 3GPP PLMNs.

The ProSe Application ID used for Open ProSe Discovery is called the Public ProSe Application ID. The geographic scope of the Public ProSe Application ID may be PLMN specific, country specific, or global. This can be signified based on the contents of the PLMN ID. If it is PLMN specific, PLMN ID corresponds to the PLMN. If it is country specific, then MNC part of the PLMN ID is wild carded by "*." If it is global, then both MNC and MCC are wild carded.

Each Public ProSe Application ID is composed of the following parts:

a. The ProSe Application ID Name is described in its entirety by a data structure characterized by different levels, for example, broad-level business category (Level 0)/business subcategory (Level 1)/business name (Level 2)/shop ID (Level 3). For the purpose of presentation, a ProSe Application ID Name is usually displayed as a string of labels in which the labels represent hierarchical levels.
b. The PLMN ID that corresponds to the PLMN that assigned the ProSe Application ID Name.

Some examples for Application ID(s):

1. PLMN-specific application ID: mcc123.mnc012.ProSeApp.USA.Transport.Train;
2. Country-specific application ID: mcc123.mnc*.ProSeApp.USA.Transport.Train;
3. Global application ID: mcc*.mnc*.ProSeApp.USA.Transport.Train.

4.8.5.7 ProSe Application Code

This is a temporary code that corresponds to the ProSe Application ID. It is assumed to have a fixed length of 184 bits. The ProSe Application Code is used by the announcing UE and monitoring UE. The announcing UE obtains it from the HPLMN ProSe Function and transmits this over PC5 interface to monitoring UE(s). The monitoring UE obtains the Discovery Filter from the HPLMN ProSe Function in order to monitor the ProSe Application Code(s).

Each ProSe Application Code is composed of the following parts:

a. A temporary identity that corresponds to the ProSe Application ID name.
b. The PLMN ID of the ProSe Function that assigned the ProSe Application Code, that is, MCC and MNC.

ProSe Application Code matching considers all parts (the temporary identity and the PLMN ID) of the ProSe Application Code. In ProSe Application Code matching, the "monitoring UE" shall consider it a full match, if both PLMN ID and temporary identity match with the corresponding contents of the Discovery Filter. It shall consider it a partial match if the PLMN ID matches fully and the temporary identity matches partially with the corresponding contents of the Discovery Filter.

A ProSe Application Code is allocated per "announcing UE" and per application and has an associated validity timer that runs both in the ProSe Function and in the UE.

The ProSe Function may decide at any time to replace a previously allocated ProSe Application Code by providing the UE with a new ProSe Application Code in which the temporary (UE specific) identity is changed. Replacing a ProSe Application Code resets the corresponding validity timer both in the ProSe Function and in the UE.

In case of Open ProSe Discovery:

- when the "announcing" UE wants to announce something, it should send a Discovery Request containing the Public ProSe Application ID to the ProSe Function, and the ProSe Function assigns a ProSe Application Code.
- when the "monitoring" UE wants to monitor something, it should send a discovery request containing the full or a subset of the Public ProSe Application ID, for example, it may provide two out of the n levels of the full Public ProSe Application ID.

Examples for ProSe Application Code(s) could be the following:

1. A PLMN-specific code: 11-0001111011-0000001100-1101111110000
   - Scope = 11 indicates that the ProSe App Code is PLMN specific and MCC, MNC are included;

- MCC = 0001111011,
- MNC = 0000001100,
- Random Temp ID = 1111111110000.

2. A country-specific code: 11-0001111011-0000000000-1101111110000.
3. A global code: 11-0000000000-0000000000-1101111110000.

Legend: The symbol "-" is used as a separator, not required by the protocol definition specified in 3GPP. This also applies to the examples provided in the subsequent sections.

4.8.5.8   ProSe Application Mask

This is used for partial matching of ProSe Application Codes received on the PC5 interface. A ProSe Application Mask is contained in a Discovery Filter. A ProSe Application Mask consists of one or more applicable parts of temporary identities of ProSe Application Codes to allow partial matching of ProSe Application Codes.

Examples for ProSe Application Mask(s) could be the following:

1. A PLMN-specific mask: 11-1111111111-1111111111-1111111111111.
2. A country-specific mask: 11-1111111111-0000000000-1101111110000.
3. A global mask: 11-0000000000-0000000000-1101111110000.

For a full match, the ProSe Application Mask should be set to all 1's. For a partial match, it is the temporary identifier portion of the ProSe Application Code that can match partially but the PLMN portion of the ProSe Application Code should match fully. Thus, the ProSe Application Mask should be set in a way that the mask is applied to the ProSe Application Code, that is, the portion of the temporary identifier that is not required to match is set to 0's, while all other bits are set to 1. The length of the ProSe Application Mask should be the same as the length of the ProSe Application Code.

4.8.5.9   Discovery Filter

This consists of ProSe Application Code(s), ProSe Application Mask(s), and a Time To Live (TTL) parameter. The length of the ProSe Application Code and the ProSe Application Mask should be the same. A TTL indicates for how long the related ProSe Application Code or ProSe Application Mask in the Discovery Filter is valid after it is received.

A Discovery Filter is provided to a monitoring UE by its HPLMN ProSe Function. It is used by the monitoring UE to selectively match ProSe Application Codes received on the PC5 interface.

The scope of ProSe Application Code and ProSe Application Mask within the Discovery filter allocated to the UE is dependent on the requested ProSe Application ID. The scope refers to country specific or global or PLMN specific. The ProSe Function is expected to allocate the Discovery Filter which contains ProSe Application Code and ProSe Application Mask(s) in the same scope as that of the requested ProSe Application ID. For instance, if the requested ProSe Application ID is PLMN specific, the ProSe Function shall allocate one or more PLMN-specific ProSe Application Code and the PLMN ID portion of the ProSe Application Mask is set to match the full PLMN ID of that specific PLMN.

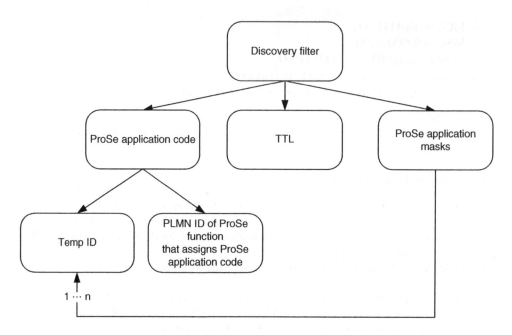

Figure 4.35   Relationship between Discovery Filter and Identifier(s)

Figure 4.35 shows the relationship between Discovery Filter, ProSe Application Code, and ProSe Application Mask(s).

4.8.5.10   EPC ProSe User ID (EPUID)

This is an identifier generated by the ProSe Function for the UE when it performs registration for EPC-level discovery.

4.8.5.11   ProSe Function ID (PFID)

This is an identifier used to identify the ProSe Function.

4.8.5.12   Application Layer User ID (ALUID)

This is an identifier designated to the user when registering with a third-party application server.

4.8.6   Illustration for Match Event

Example scenario: The Monitoring UE is interested in train services. ProSe Application ID that corresponds to Train Services is configured in an application within the UE.

1. The "monitoring" UE sends a discovery request with "monitor" command. It contains the ProSe Application ID (in this case a PLMN-specific application ID): mcc123.mnc012.ProSeApp.USA.Tranport.Train.
2. The ProSe function provides a discovery filter to the "monitoring" UE that consists of "ProSe Application Code," "ProSe Application Mask," and "TTL" to the UE:
   a. ProSe Application Code: 11-0001111011-0000001100-11110000 (Value 1)
   b. ProSe Application Mask: 11-1111111111-1111111111-11111111 (Value 2)
   c. TTL: 5 min.
3. The "announcing" UE sends discovery request with "announce" command. It contains the ProSe Application ID: mcc123.mnc012.ProSeApp.USA.Tranport.Train.
4. The ProSe function provides a ProSe Application Code to the "announcing" UE.
   a. ProSe Application Code: 11-0001111011-0000001100-11110000 (Value 3).
5. The "monitoring" UE discovers the "announcing" UE based on the ProSe Application Code and executes the match procedure.

3GPP TS 24.334 [5] defines the "Match" event as follows:

There is a match event when, for any of the masks in a Discovery Filter, the output of a bitwise AND operation between the ProSe Application Code contained in the PC5_DISCOVERY message and this mask matches the output of a bitwise AND operation between the mask and the ProSe Application Code in the same filter.

Thus, the "monitoring" UE perform bitwise "AND" operations on its own ProSe Application Code and the ProSe Application Masks. It then compares the result with the bitwise "AND" operation of the announced ProSe Application Code and ProSe Application Masks. If the comparison results in a match, then the UE can consider it as a "match event." Here we show how this operation works for the example scenario. The "monitoring" UE should perform a bitwise "AND" operation as follows:

1. Result1 = Value 1 "AND" Value 2
2. Result2 = Value 3 "AND" Value 2
3. Check for Match event: is Result 1 == Result 2?

In this example, Result 1 and Result 2 are identical, thus it is considered as a "match event." In this case, this is obvious as the ProSe Application Code matches and the content of the ProSe Application Mask is set to all 1's accordingly resulting in a "full match" event.

Example scenario: High-level scenario remains the same. The "Monitoring" UE requests for train services. The main difference is in the values allocated by the ProSe function for the ProSe Application Code and the ProSe Application Mask to the "monitoring" UE in step 2 (above).

1. PLMN-specific application ID: mcc310.mnc012.ProSeApp.USA.Transport.Train.
2. PLMN-specific ProSe Application Code known by the Monitoring UE
        11-0001111011-0000001100-11<u>011</u>100 (Value 1).
3. PLMN-specific ProSe Application Mask known by the Monitoring UE
        11-1111111111-1111111111-111<u>00000</u> (Value 2).

4. PLMN-specific ProSe Application Code provided by the announcing UE
   11-0001111011-0000001100-11**011111**0 (Value 3).

In this example scenario, the "monitoring" UE should perform a bitwise "AND" operation as follows:

1. Result1= Value 1 "AND" Value 2
2. Result2 = Value 3 "AND" Value 2
3. Check for Match event: is Result 1 == Result 2?

In this case, Result 1 and Result 2 are identical thus it is considered as a "match event." However, the temporary identifier parts of the ProSe Application Code are not identical and the match event occurred mainly as a result of using a ProSe Application Mask that is set to 1's for the parts where a match is required and 0's for the parts where it can be masked, thus resulting in a "partial match" event.

4.9 Deployment Scenarios

4.9.1 ProSe Direct Discovery

Table 4.2 shows scenarios for ProSe Direct Discovery when UE A and UE B are in-coverage or out-of-coverage of the E-UTRAN network.

Table 4.2 ProSe Direct Discovery Scenarios

#	Description	UE A	UE B	Example	Part of Release 12
1A	Out-of-coverage	Out-of-coverage	Out-of-coverage		No
1B	Partial-coverage	In-coverage	Out-of-coverage		No
1C	In-coverage-single-cell	In-coverage	In-coverage		Yes
1D	In-coverage-multi-cell	In-coverage	In-coverage		Yes
2	Relay	In-coverage UE to Network Relay	Out-of-coverage		No

Table 4.3 Direct Communication Scenarios

# Description	UE A	UE B	Example	Part of Release 12
1A Out-of-coverage	Out-of-coverage	Out-of-coverage		Yes
1B Partial-coverage	In-coverage	Out-of-coverage		Yes
1C In-coverage-single-cell	In-coverage	In-coverage		Yes
1D In-coverage-multi-cell	In-coverage	In-coverage		Yes
2 Relay	In-coverage UE to Network Relay	Out-of-coverage		No

4.9.2 ProSe Direct Communication

Table 4.3 shows scenarios for ProSe Direct Communication when UE A and UE B are in-coverage/out-of-coverage of the E-UTRAN network. When UE A has a role of "transmitting UE" and UE B has a role of "receiving UE," UE A sends messages and UE B receives them. UE A and UE B can change their transmission and reception role. The transmission from UE A can be received by one or more UEs.

4.10 Public Safety Use Cases

This section provides some background information on typical Public Safety use cases for proximity services. In addition to the term "ProSe Direct Communication," following terminology is used:

ProSe Communication: communication between two or more ProSe-enabled UEs in proximity. The term "ProSe Communication" could refer to any or all of the following:

- ProSe E-UTRA Communication between only two ProSe-enabled UEs or
- ProSe Group Communication or ProSe Broadcast Communication among Public Safety ProSe-enabled UEs or
- ProSe-assisted WLAN direct communication.

ProSe Group Communication: a one-to-many ProSe Direct Communication, between more than two Public Safety ProSe-enabled UEs in proximity, by means of a common ProSe Communication path established between the Public Safety ProSe-enabled UEs.

ProSe Broadcast Communication: a one-to-all ProSe Direct Communication, between all authorized PS ProSe-enabled UEs in proximity, by means of a common ProSe Communication Path established between these PS UEs.

Push-To-Talk (PTT) voice is the most critical means of communications for first responders in emergency situations and cannot be compromised. PTT is a feature where one person can talk at a certain time (has the "floor" to talk) while all other group members can only listen. Although the focus of applications over D2D ProSe Communication is PTT voice communications, other forms of ProSe Communication applications are also considered to be equally important.

4.10.1   Use Cases for ProSe Communication

ProSe Direct Communication means "communication in Direct Mode" and this is different from ProSe Communications via the network infrastructure as described in earlier sections.

For ProSe Group Communication or one-to-many ProSe Direct Communication involving more than two Public Safety ProSe-enabled UEs, the Incident Commander (IC) will assign team members to specific groups of users (detailed to perform a specific task) where each group is having independent ProSe Group Communication. Bifurcation of each team allows the incident commander to manage these groups more effectively and ensures that their communications are exclusive/preempted to their task. In current Public Safety LMR systems, off-network (Direct Mode) operations are often a method used for on-scene communications, in particular by the Fire Service, whether their existing trunked network is operational and has coverage in this particular area or not.

Furthermore, when a task force is manually switched to off-network (one-to-many ProSe Direct Communication) that has had one or more of its team members may move off ProSe Direct Communication-coverage (whether it was intentional or not). In such a situation, if a UE provides the user an opportunity to determine which users are in-ProSe Direct Communication-coverage at any given time, users could use this capability to determine information regarding the target user. They can determine the target user by seeing his status of availability and whether the target user is in-ProSe Direct Communication-coverage. Then, a procedure can be triggered to deterministically establish an alternative communication path to try to reach him. Additionally, UEs that are switched to ProSe Direct Communication (while being in-coverage) require continuous LTE connectivity to EPC for messages, maps, pictures, video exchanges with group communications via the infrastructure LTE network. For those UEs that are out-of-coverage, the EPC connectivity is provided by UE-to-Network Relays wherever feasible.

It is noteworthy that in a D2D (off network) environment, the incident commander could configure two users for private communication by setting up ProSe Direct Communication. Direct Communication is necessary and critical to Public Safety and this mimics the Private Call in today's LMR trunked systems. This is often communication between a supervisor and one of the people under his command that is a member of the larger group. One use case could be team leader or supervisor of a police group operating off network communicating to a sniper on the roof giving a "shoot" or "don't shoot" message at a critical moment. They may determine that immediate dialog might not be beneficial to a larger group during that critical moment. Moreover, this feature is purposely used sparingly and only specific individuals/devices are assigned to ProSe Direct Communication, such that sharing information among the group of incident members is not jeopardized.

4.10.2   Use Cases for Network to UE Relay

While the National Public Safety Broadband Network (NPSBN) will be a primary, reliable transport of Public Safety voice and data, there are many situations where voice and data communications will be required in areas where the NPSBN is not available. NPSBN Users (NPSBN-U) may be outside of the range of the network, such as first responders in a rural area assisting in a response to a plane crash or police officers inside a residence responding to a domestic issue. Thus, off-network voice communications must be immediately accessible to users in the absence of the NPSBN. This includes areas and locations where the ability to access nonterrestrial communications can be impaired such as within a building (e.g., due to steel structure) and other enclosed areas where nonterrestrial communications may not be available. Additionally, there may be times when users may wish to communicate off-network. Today, firefighters often join a local communications network, which does not leverage the fixed network, but rather relies on either direct communications between the user devices or communications via a local repeater on-scene. Firefighters can voluntarily leave the fixed network either due to the unpredictable coverage of the fixed network, or if the coverage of direct communications or the local repeater is well known, based on experience.

Thus, there will be occasions where a user may be within network coverage and will need to communicate with users who are on the network and off-network, such as an Incident Commander supporting fire response activities. These users must be able to communicate to users on the fixed network, such as dispatch, as well as the local users who are off-network or when it is desirable to provide voice, data, and video connections between users without connection to the network even if they are within network coverage.

Thus, a relay function is critical for off-network communications when NPSBN coverage is not sufficient to support the Public Safety mission. For firefighters who are responding to a wildfire while outside of the coverage of the fixed network, if one user becomes encircled by the wildfire and is beyond the range of the IC, but within the range of another device that can act as a relay, the endangered firefighter can still update his status to the IC.

4.10.3   Performance Characteristics

Performance and E-UTRAN characteristics for Public Safety related ProSe Communication are described in this section (see 3GPP TR 36.843 [14]).

4.10.3.1   Basic Operations

– With concurrent on-network operation, there may not be more than six to eight ProSe Direct Communication groups (i.e., one-to-many communication) at an incident scene.
– With concurrent on-network operations, there may not be more than 12–16 users assigned to each ProSe Direct Communication group but the group size could be expanded to 50–70 users to accommodate a search and rescue team.

- With concurrent on-network operations, two incident service members may have an authorized "private call" using ProSe Direct Communication.
- Geographic area of operations for D2D ProSe Direct Communication could be up to 1.5 mile radius per incident scene.

4.10.3.2 Coverage

- ProSe Communication for PS UEs is needed among in-coverage UEs, out-of-coverage UEs, and a mixture of UEs in and out-of-coverage.
- Determination is needed regarding within user(s) who are in ProSe Communication-coverage at any given time.
- Maintaining concurrent ProSe Communication (off-network) and LTE connectivity to EPC is required regardless of whether UEs are in-coverage or out-of-coverage. The LTE connectivity to EPC for out-of-coverage UEs are provided via UE-to-Network Relays.

4.10.3.3 Applications

- The applications expected to be supported by ProSe communication for Public Safety are voice, location, low-speed data (SMS, report/query, sensor, etc.), and pictures (optional video, if possible) with voice as the most critical means of communications.
- Emergency Alert: The alert will be sent to either the group leader or all other group members (undercover law enforcement operations, fire ground operations).
- Ability to locate team members: That is the ability for the team leader to send a query to acquire the location of an unresponsive team member. The UE (user equipment) used by the unresponsive team member, if it is functional would respond automatically, providing the location of the unresponsive team member to the team leader.

4.10.3.4 System Aspects

- There is an expectation that when a UE is served by the E-UTRA network, the network will perform radio resource allocation for Proximity Services Communication whether the Proximity Services Communication is using the same carrier or a different carrier from that used by the cellular E-UTRA network the UE is attached to.
- Ability to perform dissemination of radio resource allocations and updates within the order of seconds to meet the performance expectations of Public Safety users.

4.11 Outlook to Enhanced Proximity Services

As explained earlier, some ProSe Features were deprioritized from 3GPP Release 12, thus it is expected that these features will be specified in the subsequent release(s). In high level, following are the main features that are standardized in Release 12:

- Open ProSe Direct Discovery
- EPC-level Discovery

- EPC-assisted interworking with WLAN direct
- ProSe Direct Communication 1-many

As some features were not completed in Release 12, it is expected that the Release 13 study item [32] will address necessary enhancements in order to support the following:

a. Restricted ProSe Direct Discovery for non-Public Safety use
b. ProSe Direct Discovery for Public Safety use
c. Support for "Model B" ProSe Direct Discovery for all use cases (i.e., open and restricted)
d. Enhancements to the procedures for Open ProSe Direct Discovery such as management of ProSe Application IDs and metadata at the ProSe Function from application server over PC2 and revocation of ProSe Application Code from the ProSe Function
e. Status determination and reporting, including location status, presence status, group status, and UE network coverage status in ProSe for Public Safety use cases, if it is in scope of 3GPP standardization. If some of this status information is exchanged only as application layer signaling, then it will not be in the scope of the 3GPP work.
f. ProSe UE-Network relays for Public Safety use
g. ProSe UE-UE relays for Public Safety use
h. ProSe Direct Communication one-to-one for Public Safety use
i. Requirements for service continuity and QoS/priority/preemption of ProSe Direct Communication sessions
j. Architecture enhancements to support proximity estimations, for example, how near or how far a discovered UE is from the discovering UE. Based on that, additional ProSe discovery, range classes could be introduced.

4.12   Terms and Definitions

4.12.1   Home PLMN

It refers to the home operator/PLMN of the subscriber. It identifies the PLMN in which the subscriber's profile is stored. When the subscriber roams to other networks, subscription information is received from the HPLMN.

4.12.2   Equivalent Home PLMN

It refers to the PLMN that is "equivalent" to the HPLMN of the subscriber and is stored in a list called EHPLMN list on the USIM. An operator can provision multiple HPLMN codes by adding them to the EHPLMN list. The EHPLMN list may also contain the HPLMN code derived from the IMSI. "Equivalent" means that the PLMNs within the EHPLMN list are treated equally when the UE performs network selection.

4.12.3   Visited PLMN

It refers to the VPLMN where the user is currently roaming. The user is unable to register with the HPLMN either because of lack of coverage in the home country or because she/he is in a foreign country. For more details about Visited PLMN, please refer to the definition section of this chapter.

4.12.4   Registered (Serving) PLMN

It refers to the PLMN where user's device is currently registered. The registered (serving) PLMN is the same as the HPLMN when user's device resides in the coverage area of the HPLMN. Registered PLMN is the VPLMN when the subscriber resides outside the coverage of the HPLMN but is within the coverage area of the VPLMN.

4.12.5   Local PLMN

This is a PLMN in which the monitoring UE is authorized by the HPLMN to use radio resources to engage in ProSe Direct Discovery. The Local PLMN is not the same as the registered (serving) PLMN of the UE.

4.12.6   Hybrid Adaptive Repeat and Request

It refers to a physical layer retransmission combining mechanism. In a physical layer Hybrid Adaptive Repeat and Request (HARQ) operation, the receiver stores the packets with failed CRC checks and combines the received packet when a retransmission is received. Both soft combining with identical retransmissions and combining with incremental redundancy are facilitated [33].

4.12.7   Radio Link Control

It refers to the RLC sublayer in the LTE radio. Functions of the RLC sublayer are performed by RLC entities. The RLC entity is configured in the UE and in the eNB. For each RLC entity configured at the eNB, there is a peer RLC entity configured at the UE and vice versa.
    Function of an RLC entity is

– receive and deliver RLC Service Data Units (SDUs) from or to upper layer;
– send and receive RLC PDUs to and from its peer RLC entity via lower layers.

    An RLC entity can be configured to perform data transfer in one of the following three modes: Transparent Mode (TM), Unacknowledged Mode (UM), or Acknowledged Mode (AM). Consequently, an RLC entity is categorized as a TM RLC entity, an UM RLC entity, or an AM RLC entity depending on the mode of data transfer that the RLC entity is configured to provide. RRC is generally in control of the RLC configuration.

4.12.7.1   TM Radio Link Control (RLC TM)

In the TM mode, the RLC entity only delivers and receives the PDUs on a logical channel but does not add any headers to it, thus no track of received PDUs is kept between the receiving and transmitting entity. The TM mode of operation is only suited for services that do not use physical layer retransmissions or services that are not sensitive to the delivery order [33].

4.12.7.2 UM Radio Link Control (RLC UM)

In the UM mode of operation, the RLC entity provides more functionality, including in-sequence delivery of data that might be received out of sequence due to HARQ operation in lower layers. The UM Data (UMD) are segmented or concatenated to suitable size RLC SDUs and the UMD header is then added. The RLC UM header includes the sequence number for facilitating in-sequence delivery and duplicate detection [33].

4.12.7.3 AM Radio Link Control (RLC AM)

In the AM mode of operation, the RLC entity supports the functionalities provided in UM mode and in addition, it also supports retransmission if PDUs are lost as a result of operations in the lower layers. The AM Data (AMD) can also be resegmented to fit the physical layer resource available for retransmissions [33].

4.12.8 Logical Channel Prioritization

It refers to the Logical Channel Prioritization procedure that is applied when a new data transmission is performed. RRC controls the scheduling of uplink data by signaling for each logical channel: *priority* where an increasing *priority* value indicates a lower priority level, *prioritisedBitRate* which sets the Prioritized Bit Rate (PBR), *bucketSizeDuration* which sets the Bucket Size Duration (BSD) (see 3GPP TS 36.321 [17]).

4.12.9 System Information

It refers to the System Information (SI) such as NAS- and AS-related parameters that can be broadcasted by the network. On the basis of the characteristics and usage of parameters, it can be divided into the MIB and a number of SIBs.

4.12.9.1 Master Information Block (MIB)

The MIB includes a limited number of most essential and most frequently transmitted parameters that are needed to acquire other information from the cell and is transmitted on the Broadcast Channel (BCH) every 40 ms and is repeated within 40 ms.

4.12.9.2 System Information Block (SIB)

SIBs are transmitted in SI messages except for SIB1. SIB 1 is scheduled periodically with a fixed periodicity of 80 ms and is repeated within 80 ms. SIBs having the same scheduling requirement (periodicity) can be mapped to the same SI message. There may be multiple SI messages transmitted with the same periodicity.

4.12.10   OFDM Symbol

An OFDM symbol is a linear combination of the symbols transmitted simultaneously on the OFDM subcarriers.

4.12.11   Dual-Rx UE

A dual-Rx UE is an UE supporting two receivers. It can receive and decode two different radio signals simultaneously. For instance, it is an UE that is able to receive information from an LTE eNB and a Code Division Multiple Access (CDMA) base station at the same time.

References

[1] 3GPP TR 22.803: "Feasibility Study for Proximity Services (ProSe)".
[2] 3GPP TS 22.278: "Service Requirements for the Evolved Packet System (EPS)".
[3] 3GPP TS 23.303: "Proximity-based services (ProSe)".
[4] 3GPP TS 29.343: "Proximity-Services (Prose) Function to Proximity-Services (ProSe) Application Server Aspects (PC2); Stage 3".
[5] 3GPP TS 24.334: "Proximity-Services (Prose) User Equipment (UE) to Proximity-Services (ProSe) Function Aspects (PC3); Stage 3".
[6] 3GPP TS 29.344: "Proximity-Services (Prose) Function to Home Subscriber Server (HSS) Aspects (PC4a); Stage 3".
[7] OMA LIF TS 101 v2.0.0, Mobile Location Protocol, draft v.2.0, Location Inter-operability Forum (LIF), 2001.
[8] 3GPP TS 36.300: "Evolved Universal Terrestrial Radio Access (E-UTRA) and Evolved Universal Terrestrial Radio Access (E-UTRAN); Overall Description".
[9] 3GPP TS 29.345: "Inter-Proximity-Services (Prose) Function Signalling Aspects (PC6/PC7); Stage 3".
[10] 3GPP TS 36.413: "S1 Application Protocol (S1AP)".
[11] IETF RFC 2616: "Hypertext Transfer Protocol -- HTTP/1.1".
[12] IETF RFC 3588: "Diameter Base Protocol".
[13] 3GPP TS 36.211: "Evolved Universal Terrestrial Radio Access (E-UTRA); Physical Channels and Modulation".
[14] 3GPP TR 36.843: "Study on LTE Device to Device Proximity Services; Radio Aspects".
[15] 3GPP Tdoc R2-143672, "Introduction of ProSe", Qualcomm Incorporated et al: http://www.3gpp.org/ftp/tsg_ran/WG2_RL2/TSGR2_87/Docs/R2-143672.zip.
[16] 3GPP TS 24.301: "Non-Access-Stratum (NAS) Protocol for Evolved Packet System (EPS)".
[17] 3GPP TS 36.321: "Medium Access Control (MAC) Protocol Specification".
[18] 3GPP Tdoc R2-143733, "Consideration on in-coverage Definition", LG Electronics Inc: http://www.3gpp.org/ftp/tsg_ran/WG2_RL2/TSGR2_87/Docs/R2-143733.zip.
[19] 3GPP Tdoc R2-143572, "ProSe Multi-Carrier Support", Ericsson: http://www.3gpp.org/ftp/tsg_ran/WG2_RL2/TSGR2_87/Docs/R2-143572.zip.
[20] 3GPP draft Meeting report, R2-14xxxx_draft_report_RAN2_87_Dresden_(v0.1)_140822, "Draft Report of 3GPP TSG RAN WG2 meeting #87 Dresden, Germany, August 18–22, 2014": http://www.3gpp.org/ftp/tsg_ran/WG2_RL2/TSGR2_87/Report/R2-14xxxx_draft_report_RAN2_87_Dresden_(v0.1)_140822.zip.
[21] IETF RFC 4862: "IPv6 Stateless Address Autoconfiguration".
[22] IETF RFC 3927: "Dynamic Configuration of IPv4 Link-Local Addresses".
[23] IETF RFC 826: "An Ethernet Address Resolution Protocol".
[24] Wi-Fi P2P Specification: Wi-Fi Alliance Technical Committee P2P Task Group, "Wi-Fi Peer-to-Peer (P2P) Technical Specification", Version 1.1.

[25] IEEE Std 802.11-2012: "IEEE Standard for Information technology - Telecommunications and Information Exchange between Systems - Local and Metropolitan Area Networks - Specific Requirements - Part 11: Wireless LAN Medium Access Control (MAC) and Physical Layer (PHY) Specifications".

[26] 3GPP TS 31.102: "Characteristics of the Universal Subscriber Identity Module (USIM) Application".

[27] 3GPP TS 24.333: "Proximity-Services Management Object (MO)".

[28] 3GPP TS 33.303: "Proximity-Based Services (ProSe); Security Aspects".

[29] 3GPP TR 32.844: "Study of charging support for ProSe one-to-many Direct Communication for Public Safety Use".

[30] 3GPP TS 32.277: "Proximity Based Services (ProSe) Charging".

[31] 3GPP TS 32.299: "Telecommunication Management; Charging Management; Diameter Charging Applications".

[32] Rel-13_description_20140630: http://www.3gpp.org/ftp/Information/WORK_PLAN/Description_Releases /Rel-13_description_20140630.zip.

[33] Holma H. and Toskala A. (2009) *LTE for UMTS OFDMA and SC-FDMA Based Radio Access*. Wiley John Wiley & Sons, The Atrium, Southern Gate, Chichester, West Sussex, PO19 8SQ, UK.

5

Group Communication Over LTE

5.1 Introduction to Group Communication Services

In general, a Group Communication Service provides a fast and efficient mechanism to distribute the same content to multiple users in a controlled manner. As an example, the concept of group communication is used extensively in the operation of classical Land Mobile Radio (LMR) systems such as P.25 in the United States or TETRA in Europe and is used for, but not limited to, public safety (PS) organizations (also commercial realizations exist). At the moment, the primary use of a Group Communication Service in LMR is to enable a "Push to Talk" (PTT) function. PTT is a feature where one person can talk at a certain time (has the "floor" to talk) while all other group members can only listen. Thus, a Group Communication Service based on the 3GPP architecture, using Long-Term Evolution (LTE) radio technology, needs to enable PTT voice communication and meet expected performance requirements, for example, end-to-end latency should be less than 150 ms (source: 3GPP TS 22.468 [1]).

The service should allow different media types to be used as communication media to fulfill requirements from different users and environments they are operating in. For example, the capabilities of LTE allow for broadband communication, so an LTE-based Group Communication Service is expected to support voice, video, and in general any kind of data communication. Also, LTE allows a user to communicate with several groups in parallel, for example, using voice with one group and different streams of video or data with several other groups. A Group Communication Service should also allow a user to be a member of more than one group.

In order to provide a Group Communication Service, 3GPP defined two complementary features, one called Group Communication System Enablers for LTE (3GPP abbreviation GCSE_LTE) and the other one called Mission Critical Push to Talk over LTE (3GPP abbreviation MCPTT). Basic parts of GCSE_LTE were specified in Release 12, while work on requirements for MCPTT has only started in Release 13. The separation of two features was chosen to flexibly accommodate operational requirements on Group Communication Services that are expected to be different for various types of user groups (e.g., police or fire brigade) or may vary from country to country. The enablers defined in GCSE_LTE will provide building blocks for the application layer functionality provided by MCPTT or any other application used for group communication.

LTE for Public Safety, First Edition. Rainer Liebhart, Devaki Chandramouli, Curt Wong and Jürgen Merkel.
© 2015 John Wiley & Sons, Ltd. Published 2015 by John Wiley & Sons, Ltd.

This separation of enablers and application layer functionality was chosen to enable more flexible adaptations to regional requirements on how the group communication service has to work. It not only allows to have different service behavior between, for example, applications designed for the United States and Europe, but it would also allow for different service behavior between different usage scenarios such as police, firefighters, or private security companies.

In case other countries or user groups show interest and require specific behavior of the Group Communication Service deviating from the currently specified one, it is assumed that no changes to the GCSE_LTE enabler are required but only to the application layer functionality as defined by MCPTT or other type of new applications that uses the capabilities offered by GCSE_LTE. In this case, a new service can be deployed more rapidly as no changes to core and radio network components are needed.

5.2   Group Communication System Enablers for LTE

GCSE_LTE or briefly GCSE were defined by 3GPP in Release 12 to mimic the "PTT" transport behavior of existing legacy systems for Public Safety (PS) group communication such as TETRA or P.25 on a system based on LTE radio technology. Unlike legacy systems GCSE not only allows delivering voice in a resource efficient way to a group of users spread over a large geographical area but also delivering video or any data in general.

GCSE allows devices (in 3GPP terminology User Equipments or UE(s)) to participate in group communication for several groups in parallel, which can be one or several voice-, video-, or data-communications. It is up to the application residing in the device and its counterpart in an application server in the network called Group Call System Application Server (GCS AS) to determine on how these parallel communications will be handled. In other words, GCSE only provides lower layer functions such as the efficient delivery of voice and data to a group of UEs at the same time. The application layer handles group interaction such as floor control that decides which one of the UEs that raised a communication request is allowed to communicate (i.e., send voice, video, or data) next or mechanisms to join or leave an ongoing group communication and create, change, or delete groups (the so-called group management).

A UE can be a member of many groups out of which it can select some that are currently of interest. This could be done by the user via a user interface or remotely by application layer functionality residing in the network.

GCSE allows defining groups with a limited geographical scope and a group member will only be able to participate in group communication when being located within this geographical area.

As mentioned earlier, the UE can receive different media at the same time and also transmit different media at the same time. One use case for such a scenario could be a firefighter that receives voice and also transmits voice on group A, while his helmet camera transmits video also to group A and, even likely, to another group B and perhaps vital functions such as heart rate or breathing frequency to a group C. It is up to the application layer to decide if group A is already having an ongoing voice communication and whether a parallel setup of additional communication media to the same group A is deemed useful – the enablers would allow for this. In this case most likely an additional voice communication does not make sense, but video communication might be beneficial provided the destination device is able to display the video and the situation allows looking at the video screen. This is why the decision is left to the application layer and, in most cases, to the user ultimately.

GCSE also provides priority and preemption mechanisms so that group communication traffic with high priority will be served before other types of traffic with lower priority. Furthermore, in case of congestion, GCSE allows for preempting an ongoing traffic with lower priority to allow a group communication request with higher priority to succeed.

5.3   Principles of Group Communication over LTE

GCSE provides two types of delivery mechanisms for downlink traffic between the GCS AS and UEs. The first type of delivery mechanism is called Unicast Delivery. It uses Evolved Packet System (EPS) bearers (see Section 1.6.6) between GCS AS and UE for distributing media. The second type of delivery mechanism is called Multimedia Broadcast/Multicast Service (MBMS) Delivery. It uses a broadcast bearer of an MBMS bearer service to deliver the media to many UEs in a point-to-multipoint manner. Please refer to Section 1.12, for a description of the MBMS feature.

For uplink media transmission, the UE uses a "normal" EPS bearer to send data to the GCS AS.

Figure 5.1 shows the overall GCSE architecture showing functional blocks for Unicast Delivery and MBMS Delivery. The GCS AS uses the GCSE capabilities to provide group communication among group members. The GCS AS also performs group management-related functions such as membership control, floor control, security (including authentication and authorization of group members, protection of data transmission), and any other functions needed for group management. Group communication dispatcher function may also be implemented within GCS AS. Functions within the GCS AS (i.e., application domain) are not specified in Release 12.

While the EPS can be operated and maintained by a network operator the GCS AS can be operated and maintained by third party distinct from the network operator; for example, by a PS agency such as police or fire department. This allows for a maximum flexibility in deploying and operating the service. The GCS AS is only required to adapt to standardized interfaces such as MB2, Rx, and SGi.

As explained in Chapter 1, MBMS is a feature that was introduced for LTE in Release 9. In order to enable MBMS Delivery for GCSE, two new interfaces were specified. With the help of these interfaces, the application (e.g., MCPTT) makes use of the GCSE enabler to deliver its content to many users at the same time. MB2 is a new reference point with control and user plane part standardized as a secure third-party interface to utilize the MBMS delivery mechanism. Another new interface is called GC1, which is between the application in the UE and the GCS AS to allow application level control signaling such as group management and floor control, and for relaying any MBMS-specific bearer configuration data received from the BM-SC. Unlike MB2, GC1 will not be specified in 3GPP Release 12 and is expected to be specified in Release 13 as part of MCPTT feature. For solutions based on 3GPP Release 12, GC1 signaling is assumed to be proprietary and is carried over the user plane (i.e., EPS bearers) similar to Smartphone applications today. From EPS perspective GC1 is an "over-the-top" type of signaling connection.

For Unicast Delivery, the existing SGi and Rx interfaces as described in Chapter 1 are reused. The Rx interface allows the GCS AS to act as an Application Function (AF) as defined in PCC architecture (see chapter 1.6.7) to interwork with the EPS via Policy and Charging Rules

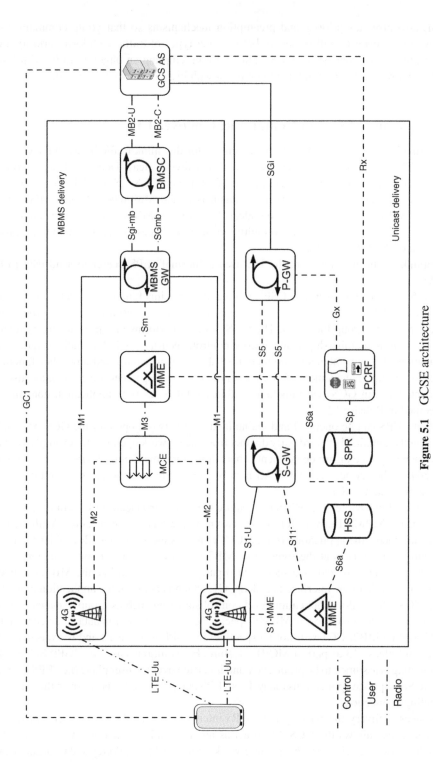

Figure 5.1 GCSE architecture

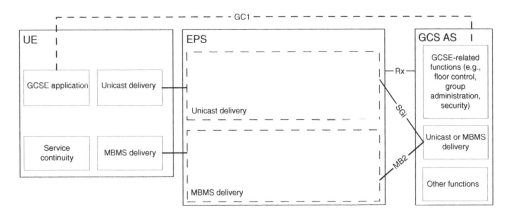

Figure 5.2 GCSE functional elements

Function (PCRF) for unicast EPS bearer creation. PCRF also receives the service characteristics (e.g., bandwidth and priority level) from the GCS AS. However, the actual priority level to be used in the EPS is determined by the PCRF, based on its operator policy. The SGi interface provides the IP connectivity between UE and GCS AS for media transfer and GC1 signaling.

The statistical traffic pattern for group communication is different from that of normal voice calls as there could be long silence periods between two short talk bursts. In LTE, it is possible to save energy (i.e., battery lifetime) in the UE with a feature called "Discontinuous Reception (DRX) mode in connected state." A similar concept exists for idle UEs, the "DRX mode in idle state." Although the UE is connected to the network (i.e., a Radio Resource Control (RRC) connection exists), there are often moments where no data are sent or received. In these moments the receiver in the UE can be switched off. The connected state DRX feature allows waking up and shutting down the receiver in the UE periodically and thus saves UE energy. The longer the DRX period, the more energy can be saved; however, the chance to lose media (e.g., voice samples) at the beginning of a talk burst is higher. On the other hand, the shorter the DRX period is, the more energy a UE will consume while wasting radio resources during long silence periods. As a consequence, to allow for fast responsiveness while still maintaining a good battery lifetime, the Evolved NodeB (eNB) needs to carefully choose DRX period and other connection-related parameters that are appropriate for the application (e.g, PTT) that uses the GCSE. This is possible with special QoS Class Identifier (QCI) values used in the context of Group Call System Enablers (GCSE) to enable the eNB to set these parameters properly.

The decision to use Unicast Delivery or MBMS Delivery is taken by the GCS AS. If the UEs report the cell they are camping on to the GCS AS, this decision can be made based on the number of UEs in a certain cell or in a larger area. With MB2, Rx, SGi, and GC1 interfaces, the GCS AS can decide and control how media related to the Group Communication Service are delivered. As the UE can potentially receive group communication-related media via Unicast Delivery or MBMS Delivery and could move in or out of a MBMS service area, 3GPP defined a service continuity mechanism allowing for seamless transition between the two delivery methods.

Figure 5.2 summarizes the new functional elements in UE and network introduced for GCSE.

Security requirements and procedures for GCSE are defined for the MB2 interface to ensure the connection between the GCS AS and BM-SC is secured and trusted. No additional security is defined for Unicast Delivery, which means the existing EPS security is reused (see Section 1.6.11). The encryption mechanism for the user data between UE and GCS AS (e.g., which security schemes are used and how session keys are distributed) is determined exclusively by the GCSE related applications and is therefore transparent to the EPS. As a consequence, the standardized MBMS security solution as specified in 3GPP TS 33.246 [2] is not used for GCSE.

No new charging functions for GCSE have been defined. Thus, the existing MBMS charging requirements as specified in 3GPP TS 23.246 [3] and 3GPP TS 32.273 [4] are reused. Namely, the BM-SC can charge the bearer service based on information such as QoS, service area, session duration, and volume of data. It is expected that subscriber level charging within the EPS is not used in case of GCSE as it would require that the EPS (i.e., the BM-SC) is able to identify single users uniquely. This cannot be ensured without using the full MBMS security solution. For Unicast Delivery, the existing EPS charging requirements (explained in Chapter 1) is reused.

5.4 Functional Entities

This section provides a description of functional entities required to support GCSE over LTE. This is in addition to the functions and functional entities required to support LTE/System Architecture Evolution (SAE) as described in Chapter 1.

5.4.1 User Equipment

A group communication capable UE has to support the following functions in addition to a normal UE:

- GC1 application layer signaling (proprietary in Release 12) with the GCS AS.
- Receiving data from a GCS AS using either Unicast Delivery or MBMS delivery or both simultaneously.
- Transmitting data in uplink toward the GCS AS using unicast EPS bearers.
- Ensure service continuity between Unicast Delivery and MBMS Delivery.
- New QCI values defined for MCPTT signaling, voice, and other media as specified in 3GPP TS 23.203 [5].

5.4.2 GCS AS

The GCS AS supports the following functions:

- GC1 application layer signaling (proprietary in Release 12) with the UE.
- Delivery of downlink data to a group of UEs either using Unicast Delivery or MBMS delivery or both simultaneously.
- Receiving uplink data from the UE via unicast bearers.
- Maintaining sessions on EPS level (e.g., establishing a unicast bearer) via Rx interface.
- Control of service continuity between Unicast Delivery and MBMS Delivery for a UE.

5.4.3   BM-SC

In addition to supporting the functions as described in Section 1.12, the BM-SC needs to support the following additional functions:

- Providing Temporary Mobile Group Identity (TMGI) and other MBMS related information (e.g, service area and frequency info) to the GCS AS via MB2 interface.
- Activation, deactivation, and modification of MBMS bearers over MB2 interface.

5.4.4   eNB, MME, S-GW, P-GW, PCRF

These functional entities have to support bearers with the new QCI values specified for MCPTT signaling, voice, and other media.

5.5   Interfaces and Protocols

5.5.1   MB2 Interface

MB2 is a new third-party secure interface standardized in 3GPP Release 12. It carries both control and user plane data and provides a standardized way for external entity (e.g., the GCS AS) to connect to the BM-SC. Such a standardized interface is not available in the MBMS architecture before Release 12.

From PS perspective, the Content Provider in Figure 5.3 is just the GCS AS. In Release 12 the decomposition of the GCS AS is not specified but the implementation of such a function can be separated into different entities (with proprietary interfaces where the server part can use H.248 protocol to control the media plane part) to carry out the media and control plane separately.

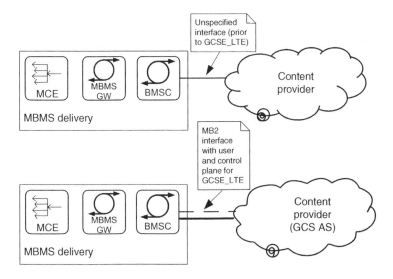

Figure 5.3   MB2 interface

When group members are dispersed with subscriptions from different LTE network operators, then the GCS AS must maintain a MB2 interface with control and user plane part to each of these networks to provide a Group Communication Service using MBMS Delivery. There is no sharing of MBMS broadcast bearers between different Public Land Mobile Networks (PLMNs), even though they are carrying the same media and might be broadcasted over the same geographical area.

Within a PLMN one single Broadcast Multicast Service Center (BM-SC) may be connected to multiple GCS AS with multiple MB2 interfaces to allow for group communication for different groups. In Release 12 the merging of two separate groups into a single group is not specified and assumed to be based on GCS AS implementation. As a consequence any inter-AS interface to allow merging of two groups is not specified in Release 12.

Figure 5.4 illustrates different possible MB2 connectivity scenarios between a GCS AS and two PLMNs offering MBMS Delivery for GCSE. In this figure the GCS AS is decomposed into a group server and a media gateway component.

In this configuration example, Group Server #1 with the assigned Media Gateway #1 has connections to the BM-SCs in PLMN 1 and PLMN 2, that is, Group Server #1 can use PLMN 1 and 2 for MBMS Delivery, while Group Server #2 has only a connection to the BM-SC in PLMN 2

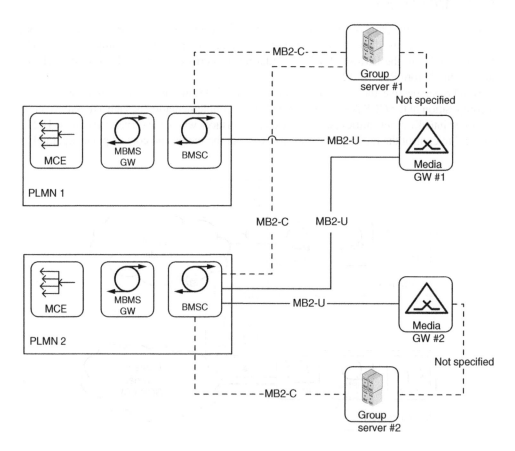

Figure 5.4   MB2 connectivity scenarios

and thus, can only use PLMN 2 for MBMS Delivery. The control plane and media plane parts for each MB2 interface have different terminating end points to allow for separation of user and control plane. When MBMS Delivery service is used for group communication like MCPTT, it is always destined to a particular MB2 interface for the lifetime of this MBMS Delivery service. Therefore, each MB2 interface is distinct from other MB2 interfaces and, for example, load sharing between multiple MB2 instances is not possible (i.e., a GCS AS cannot send different chunks of data from the same group communication via different MB2 interfaces).

The details for the MB2 interface including MB2 procedures and used protocols are specified in 3GPP TS 29.468 [6]. If the MB2 interface is an interdomain interface, it can be protected via IPSec or other standard security protocols like Datagram Transport Layer Security (DTLS). Information required to establish an IPSec security association for the MB2-U interface (e.g., IP address and port numbers) can be exchanged via the protected MB2-C interface.

The control plane (MB2-C) is based on the DIAMETER protocol as specified in IETF RFC 3588 [7] and transported over SCTP or TCP. Figure 5.5 depicts the MB2-C protocol stack. Shaded protocol layers indicate protocols that are reused without 3GPP-specific changes. The

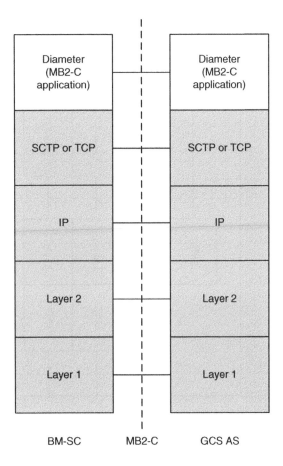

Figure 5.5   MB2-C protocol stack

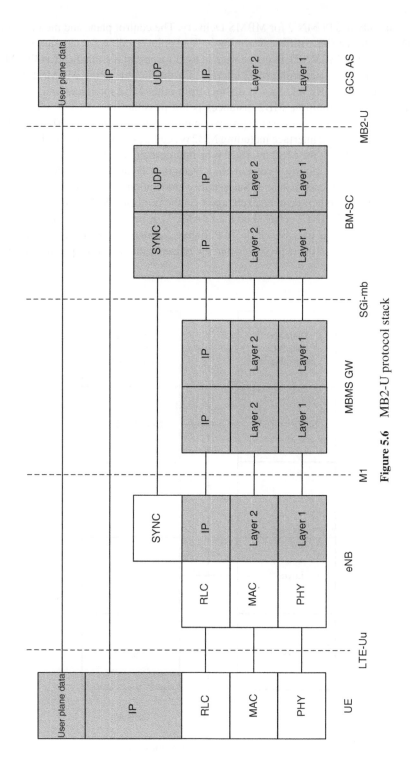

Figure 5.6 MB2-U protocol stack

BM-SC acts as a DIAMETER server acting upon requests from the client and sending notifications, and the GCS AS acts as a DIAMETER client. GCS AS can select a BM-SC based on the configuration data or DIAMETER routing principles as described in IETF RFC 3588 [7] and can be used to route DIAMETER requests to the correct BM-SC.

The user plane (MB2-U) carries user data over UDP. Figure 5.6 shows the MB2-U protocol stack and its relation to SGi-mb, M1, and LTE-Uu providing the user plane interface for MBMS Delivery toward the UE. User plane data are transported via IP. The user plane data are transported between GCS AS and UE. The BM-SC forwards the data between these protocol layers transparently. For simplicity, Layer 4 above IP is not shown explicitly in the figure. The different protocol stacks are described in 3GPP TS 29.061 [9] and 3GPP TS 36.300 [10].

The destination address for user plane data over MB2-U is determined by the BM-SC and sent to the GCS AS over MB2-C. When the BM-SC receives a MBMS bearer allocation request from the GCS AS via MB2-C, it selects an IP address for the user plane and a unique UDP port value for this IP address and provides both in its response to the GCS AS. The BM-SC transparently forwards any payload data within UPD packets to the allocated MBMS bearer.

A DIAMETER application uses normally a session-based type of transaction, that is, one long-lived DIAMETER session is created for a particular transport or bearer service. In other words, the MB2-C application could use one DIAMETER session per MBMS bearer. However, this does not fit into the signaling procedure defined for GCSE where multiple MBMS bearers identified by multiple TMGI values can be allocated and deallocated at the same time and also a notification message from BM-SC to GCS AS can combine the status of many MBMS bearers. Hence, one DIAMETER session identification cannot be used to identify a particular DIAMETER transaction. For these reasons, the MB2-C application was designed as a short-lived session consisting of a single request and answer pair, terminated after each request and answer pair interaction. The request message can group multiple DIAMETER Attribute-Value-Pairs (AVPs) together to allow a combination of requests or notifications for multiple MBMS bearers.

Table 5.1 lists Command Codes (i.e., the requests and responses that can be sent) for the MB2-C interface and their corresponding functions. For a detailed description of the procedures, see also Sections 5.7 and 5.8.

5.5.2   Rx and SGi Interfaces

In Sections 1.6.6 and 1.6.7, we described the basic concept of PCC, session management, and Quality of Service (QoS) for an EPS bearer. For group communication, the same bearer-related concepts apply in case of Unicast Delivery (i.e., user plane connection in the downlink), for uplink user plane connections to the GCS AS (e.g., talker's speech), and for the signaling exchange over GC1 interface. All these user plane data are transmitted over normal point-to-point EPS bearers. The SGi interface as described in 3GPP TS 29.061[9] provides the UE with a user plane connection to the GCS AS. This connection can be used for transfer of application layer signaling and media data. In 3GPP terminology this corresponds to a PDN connection where the PDN can be defined for a certain group like "Firemen" or a service like MCPTT. The GCS AS indicates over Rx the priority level and the GCS identifier within the service authorization request to the PCRF. The priority level and the GCS identifier are configured internally within the GCS AS. The PCRF uses these values to determine appropriate

Table 5.1 MB2-C command-code values

Command-Name	Functions
GCS-Action-Request (GAR)	The GAR command allows the GCS AS to request a BM-SC to allocate/deallocate a TMGI or to activate/modify/deactivate MBMS Delivery (i.e., a MBMS bearer corresponding to an allocated TMGI)
GCS-Action-Answer (GAA)	The GAA command allows the BM-SC to respond to the GCS AS request due to GAR
GCS-Notification-Request (GNR)	The GNR command allows the BM-SC to send a status notification to the GCS AS about the allocated TMGIs or MBMS bearer in question
GCS-Notification-Answer (GNA)	The GNA command allows the GCS AS to respond to the BM-SC due to GNR

values for QCI (new values were defined for mission and nonmission critical voice and data), ARP priority level, preemption capability, and preemption vulnerability of the corresponding EPS bearer. The PCRF also ensures that the default bearer is at least maintaining the same priority level as the EPS bearer being established for UL or DL media delivery in order to ensure the session is not aborted during a handover procedure due to possible lower priority of the default bearer.

When the UE is roaming, GCS AS can contact either the H-PCRF (PCRF in home network) or V-PCRF (PCRF in visited/roaming network) based on the existing procedure defined in 3GPP TS 23.203 [5]. It is assumed that the GCS AS receives UE's assigned IP address, Home Public Land Mobile Network (HPLMN) ID and Visited Public Land Mobile Network (VPLMN) ID via GC1 signaling. On the basis of received IP address and serving PLMN information (either HPLMN ID or VPLMN ID), the GCS AS decides if the UE is roaming and which H-PCRF has to be contacted. In the roaming case, the H-PCRF contacts a V-PCRF via the S9 interface to convey QoS information, or GCS AS connects to V-PCRF via Rx directly based on agreements with HPLMN/VPLMN operators.

5.5.2.1 Summary of Reference Points and Protocols

Table 5.2 summarizes the reference points and protocols used for GCSE:

Table 5.2 Reference points for GCSE

Reference point	Protocols	Specification
GC1	Not specified in 3GPP Release 12	Not specified in 3GPP Release 12
MB2	DIAMETER	TS 29.468 [6]

5.6   GCSE Functions

5.6.1   Unicast Delivery

Unicast Delivery in group communication refers to the reuse of normal EPS bearers for communication between device (UE) and server (GCS AS). The concept of EPS bearers has been described in Section 1.6.6. In a more general term, it is an IP pipe between two end points providing a specific QoS such as packet delay budget, admission, and scheduling priority. QoS is very important for GCSE in order to allow building a high-performance system with low call setup times.

According to 3GPP TS 22.468 [1], latency for end-to-end GCS media delivery should be less than or equal to 150 ms. When examining the existing defined QCI values defined in 3GPP TS 23.203 [5], QCI-3 with 50 ms delay budget would meet and exceed this requirement as 50 ms is less than "150 ms/2" = 75 ms. However, QCI-3 does not have high enough scheduling priority in E-UTRAN (priority given by the radio resource scheduler in eNB to packets of a QCI-3 bearer), which means that the probability of successful delivery during system congestion is not sufficiently high. Furthermore, when looking from a radio utilization standpoint, using a normal GBR bearer for group communication, for example, PTT traffic would require eNB to reserve radio resources unnecessarily most of the time as this traffic pattern is made up of short talk bursts followed by long silence periods. To overcome this drawback, 3GPP defined a new set of QCI values for GCSE. Table 5.3 extracted from 3GPP TS 23.203 [5] (Table 6.1.7) presents these new values and characteristics.

Values QCI-65 and QCI-66 were defined for GBR bearers. Both are associated with NOTE 8 in Table 5.3, in which 320 ms is mentioned as the value for the DRX cycle. DRX is an important feature to reduce UE battery consumption by temporary shutdown of the UE receiver for a given duration (the DRX cycle). If the UE is in idle mode, the shortest allowed paging cycle according to 3GPP TS 36.331 [11] is 320 ms. If, for example, a UE that is part of a GCS group needs to be paged, the network can expect that the UE will listen to the paging channel every 320 ms. After paging response, the UE will have to trigger the EPS bearer setup to communicate with the server (GCS AS). When designing the group communication service, this delay should be taken into account, if the receiving group member is assumed to be idle. The DRX value for IDLE mode can be allocated by the GCS application within the UE. It should also be noted that the actual DRX value used for paging depends on the DRX value broadcasted by E-UTRAN. According to 3GPP TS 36 304 [12], this value is determined by the shortest of the UE-specific DRX values that are allocated by the upper layers (e.g., the GCS application) and a default DRX value broadcasted in the network. If no UE-specific DRX value is configured, the default value applies. The DRX value to be used for paging is indicated by the UE to the core network (MME, Mobility Management Entity) in the Attach Request and Tracking Area Update Request.

In other group communication setup scenarios, it can be expected that the receiving group members (i.e., UEs) are mostly in connected mode. In RRC Connected mode, the shortest DRX cycle allowed by 3GPP TS 36.331 [11] is 10 ms while the longest is 2.56 s. The value to be used is determined by the eNB. As group communication services such as push-to-talk have long silence periods between speech bursts, setting the DRX cycle to 10 ms will effectively waste the UE battery power during these periods. Setting it to 2.56 s will cause the UE to miss the first portions of the speech burst. Hence, using a new QCI value will allow the eNB to set the

Table 5.3 QCI values defined for GCSE

QCI	Resource type	Priority level	Packet delay budget	Packet error loss Rate (NOTE 2)	Example services
65 (NOTE 9)	GBR	0.7	75 ms (NOTE 7, NOTE 8)	10^{-2}	Mission Critical user plane Push To Talk voice (e.g., MCPTT voice)
66		2	100 ms (NOTE 1, NOTE 8)	10^{-2}	Non-Mission-Critical user plane Push To Talk voice
69 (NOTE 9)	Non-GBR	0.5	60 ms (NOTE 7)	10^{-6}	Mission Critical delay sensitive signaling (e.g., MCPTT signaling)
70		5.5	200 ms (NOTE 7)	10^{-6}	Mission Critical Data

NOTE 1: A delay of 20 ms for the delay between a PCEF and a radio base station should be subtracted from a given PDB to derive the packet delay budget that applies to the radio interface. This delay is the average between the case where the PCEF is located "close" to the radio base station (roughly 10 ms) and the case where the PCEF is located "far" from the radio base station, for example, in case of roaming with home-routed traffic (the one-way packet delay between Europe and the US west coast is roughly 50 ms). The average takes into account that roaming is a less typical scenario. It is expected that subtracting this average delay of 20 ms from a given PDB will lead to desired end-to-end performance in most typical cases. Also, note that the PDB defines an upper bound. Actual packet delays – in particular for GBR traffic – should typically be lower than the PDB specified for a QCI as long as the UE has sufficient radio channel quality.

NOTE 7: For Mission Critical services, it may be assumed that the PCEF is located "close" to the radio base station (roughly 10 ms) and is not normally used in a long distance, home-routed roaming situation. Hence delay of 10 ms for the delay between a PCEF and a radio base station should be subtracted from this PDB to derive the packet delay budget that applies to the radio interface.

NOTE 8: In order to permit reasonable battery saving (DRX) techniques in both RRC Idle and RRC Connected mode, for the first packet(s) in a downlink talk or signaling burst, the PDB requirement is relaxed but not to a value greater than 320 ms.

NOTE 9: It is expected that QCI-65 and QCI-69 are used together to provide Mission Critical Push to Talk service (e.g., QCI-5 is not used for signaling for the bearer that utilizes QCI-65 as user plane bearer). It is expected that the amount of traffic per UE will be similar or less compared to the IMS signaling.

DRX cycle no longer than 320 ms, ensuring the same decent performance as in RRC Idle mode while still preserving decent battery power for group communication services such as PTT.

QCI-65 has higher scheduling priority than QCI-66, which means that during congestion situations, the eNB will try to schedule the resources for delivering the packets of a QCI-65 bearer before packets of a QCI-66 bearer. Thus, QCI-65 is being labeled as a "Mission Critical user plane" QCI. For the "Mission Critical" signaling counterpart, the new QCI-69 value with even higher scheduling priority is introduced. The design is analogous to IMS where the IMS signaling bearer and user plane bearer are separated and IMS signaling has higher priority compared to the corresponding IMS user plane. Although GCSE does not mandate any particular application layer technology (e.g., IMS), the signaling/user plane separation combined with different priority levels will allow great flexibility in designing group communication services based on GCSE.

The new QCI-70 value is introduced with a higher scheduling priority than the existing QCI-6/7/8/9 values as mission critical data services will have a higher priority to be transmitted than regular commercial services during congestion.

5.6.2 MBMS Delivery

MBMS Delivery uses the broadcast mode of 3GPP's MBMS feature for delivery of downlink group communication media to many users at the same time. As stated in Chapter 1, multicast mode of MBMS for delivery of downlink data is not supported in LTE, but only broadcast mode. The basic MBMS Delivery mechanisms are described in Section 1.12.

When a cell starts to support MBMS Delivery, the cell broadcasts MBMS-related information to all UEs over the Broadcast Control Channel (BCCH). The so-called BCCH modification period is calculated by multiplying the default DRX cycle (smallest value is 320 ms) with a coefficient, which is 2, 4, 8, or 16. For details please refer to 3GPP TS 36.331 [11]. Thus, the smallest BCCH modification period is 640 ms and it may take up to one or two periods before the UE notices that the cell is starting MBMS transmission. Upon receiving the notification about MBMS support, the UE has to gather first MBMS configuration data that are broadcasted by the eNB. These data allow the UE to monitor the MCCH. The TMGI value that is allocated to the MBMS bearer and provided to the UE via application layer signaling by the GCS AS is indicated to the UE via the MCCH. MCCH information is repeated with a time period between 320 ms and 2560 ms and the information can only be modified at either 5.12 or 10.24 s, which is the so-called MCCH modification period. Hence, the UE must read the MCCH information at every MCCH modification period in order to determine, if MBMS sessions for the TMGIs it is interested in are active in the MBMS Single Frequency Network (MBSFN) area. This means when the GCS AS decides to activate MBMS Delivery, it needs to take the MCCH modification period into account to determine the actual time when the UE could be aware of this delivery in addition to the possible BCCH modification period (i.e., when the cell is switched to MBMS Delivery). When broadcast for the TMGIs is activated, the media related to that TMGI value could be sent at every MCH Scheduling period (MSP), which is between 40 ms and 10.24 s. For mission critical PTT media, the shortest cycle (i.e., 40 ms) is used. Hence, the UE will wake up every scheduling period (e.g., every 40 ms) to read the Mobile Subscriber Identity (MSI) in order to determine whether the corresponding TMGIs are scheduled during this period for media delivery. If the UE is aware that MBMS Delivery for the indicated TMGIs is ongoing, it can receive data via the MTCH. These data might be encrypted by the application layer.

A PS user may be a member of multiple different (talk or video) groups, each group can be linked to a different TMGI value. Therefore, it is important that the UE can support more than one MBMS bearer service at a time. Although 3GPP TS 36.331 [11] mandates support of only one MBMS bearer service (i.e., one TMGI and one MTCH) at a time for the UE, the UE can support more MBMS bearer services, especially if they are transmitted on the same radio carrier by the serving cell. From protocol perspective, MBMS allows maintaining up to 3480 bearer services (i.e., MBMS sessions with its own TMGI) in parallel.

If the radio subframes reserved for MBMS bearers is not carrying any data (i.e., no broadcast is ongoing), they can be reused for unicast traffic. However, this is only possible for UEs that are capable to use Transmission Mode-9 (Release 10 UEs) or Transmission Mode-10 (Release 11 UEs) and are not configured to receive the MBMS bearer service corresponding to the given subframes. All prerelease 10 UEs are not able to receive Unicast Delivery in radio subframes allocated for MBMS bearer services. The Transmission Mode-9 and Transmission Mode-10 capabilities are indicated by the UE to the eNB.

According to the study documented in 3GPP TR 36.868 [13], the one-way delay of the media to reach the UE from BM-SC will require 130 ms: 40 ms from BM-SC to eNB plus in 80 ms (MSP) for the UE to read the MSI plus 10 ms from eNB to UE. After the completion of this study, 3GPP has further reduced the MSP to 40ms, thus the overall additional delay of this portion of the system is reduced to 90ms.

5.6.3   Service Continuity

Group communication media can be delivered via MBMS Delivery or Unicast Delivery. The UE uses MBMS Delivery whenever it is available. However, MBMS Delivery may become unavailable because UE moves out of the MBMS service area or due to the fact that the existing MBMS bearer is preempted by a higher priority bearer. Service Continuity allows the UE to continue to receive group communication media from the GCS AS whenever MBMS Delivery becomes unavailable.

3GPP has defined a UE-based service continuity mechanism in Release 12. As a consequence it is UE's responsibility (in a proprietary manner) to determine when to switch between MBMS Delivery and Unicast Delivery. In Release 12 the LTE radio network is not enhanced to provide any special indication to the UE, if MBMS Delivery is about to be aborted or the UE is moving away from the MBMS service area; seamless service continuity is highly dependent on the UE and application layer implementation. Basically the UE needs to determine if radio signal strength and quality are sufficient to continue using MBMS Delivery. If this is not the case, the UE informs the GCS AS via GC1 signaling that it wishes to establish a unicast bearer for Unicast Delivery. The GCS AS triggers establishment of EPS bearers appropriate for group media delivery over the Rx interface. During this process, a UE may temporarily receive the same group communication media via both MBMS Delivery and Unicast Delivery and it is up to the UE to discard duplicated data. This scenario or procedure is called "Make before Break" as a unicast bearer is established before the UE is unable to receive data via MBMS Delivery anymore.

A "Break before Make" scenario or procedure is assumed when the reception of data via MBMS Delivery is lost before Unicast Delivery is successfully established. If group communication media are delivered during this time, the loss of media will be perceived by the user.

Figure 5.7 shows the two service continuity scenarios.

A MBSFN reserved cell is a cell within or at the edge of a MBSFN area, which does neither advertise nor contribute to the MBSFN transmission but transmits other data with low power not to interfere with the MBMS signal. In case the UE moves into a MBSFN reserved cell, it can recognize that it is about to leave the MBSFN area as there is no MBSFN advertisement in the reserved cell while it still receives the MBMS signal from the neighboring cells. Thus, the UE can immediately request Unicast Delivery service from the GCS AS. In this case a "Make before Break" scenario is likely to be achieved and no interruption may occur during the switch from MBMS Delivery to Unicast Delivery.

If MBSFN reserved cells are not used, the UE can at least observe poor MBSFN signal quality (i.e., poor signal quality of the Multicast Traffic Channel (MTCH)) before leaving the MBSFN area. When to trigger the establishment of Unicast Delivery is a UE implementation

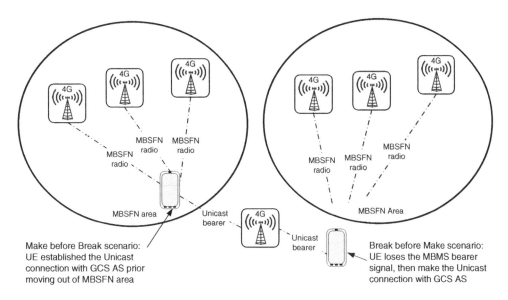

Figure 5.7   Service continuity scenarios

specific decision. In Release 12 no standardized trigger to assist the UE in switching from MBMS Delivery to Unicast Delivery exists.

To switch from Unicast Delivery to MBMS Delivery, the UE monitors the serving cell to determine if MBMS Delivery is available. This is done by existing MBMS procedures. The UE monitors the Multicast Control Channel (MCCH) to determine, if the TMGI that it is interested in (i.e., that was provided by the GCS AS via application layer signaling) is available. If it is available, the UE will use MBMS Delivery to receive downlink transmission and indicate to GCS AS that Unicast Delivery can be released. As explained earlier, reading the TMGI by a particular UE heavily depends on the BCCH and MCCH modification period settings. Thus, if a GCS AS has started MBMS Delivery, the time taken for the UE to switch from Unicast Delivery to MBMS Delivery also depends on the setting of these parameters.

5.6.4   Priority and Preemption

When a system is congested, either at radio and core network level, priority and preemption indications become important. In case of GCSE these indications are used by the network to give higher priority to selected group communication (i.e., MBMS Delivery) or to communication of selected individual group members (i.e., Unicast Delivery). There is no guarantee that resources for high-priority group communication can be allocated during a congestion situation but the probability to get resources is higher for the selected group members than for other users.

The assignment of a priority and preemption level for each group communication and its particular members is under responsibility of the GCS AS. The GCS AS assigns an Allocation Retention Priority (ARP) indication for the MBMS Delivery. This indication includes the priority level (values from 1 to 15, where 1 is the highest value), a preemption capability (i.e., whether preempting other communication is allowed), and a preemption vulnerability (i.e., whether other traffic is allowed to preempt this communication). However, this should not be confused with the ARP that is being used for EPS bearers within the network. The ARP being assigned by the GCS AS can be reassigned (or remapped) to a different ARP indication that is actually applied by the BM-SC to a MBMS bearer. For Unicast Delivery, the GCS AS uses the existing Rx interface to indicate the priority level and service identifier to PCRF as part of the service authorization request. The PCRF derives the network level ARP using existing procedure as defined in 3GPP TS 23.203 [5], for both the default and dedicated EPS bearers.

If changes to the priority level or preemption characteristics are required, GCS AS provides an update of the new service characteristics toward the PCRF via the Rx interface for Unicast Delivery. If the priority of all members within a group is being upgraded or downgraded, the GCS AS will need to treat each member separately toward the PCRF by sending individual requests to update the service characteristics. In the case of MBMS Delivery, GCS AS sends an updated priority level and/or preemption characteristics to the BM-SC (using ARP) for the MBMS Delivery session that is identified by TMGI and Flow ID. From Release 12 onwards, the network level ARP can be modified via the MBMS Session Update procedure according to 3GPP TS 23.246 [3] (explained in chapter 7). Figure 5.8 shows the overall procedure for change of ARP in case of MBMS Delivery. As the priority indication has changed, the MCE may need to change the MBMS radio allocation among the different MBMS bearers.

Modifying the priority indication in case of MBMS Delivery may cause preemption of MBMS bearers. For example, some MBMS bearers may have lower priority and may be suspended by the MCE due to radio resource congestion. If an MBMS Delivery session is modified to an even lower priority than the suspended MBMS bearers and preemption is allowed, then MCE can start to reallocate radio resources to the higher priority suspended MBMS bearers and suspend the downgraded MBMS Delivery session. An MBMS bearer is always a GBR bearer. If the resources of one of the MBSFN cells in the MBSFN area is not sufficient to meet the GBR requirements, the MCE can suspend the MBMS Delivery for the whole MBSFN area

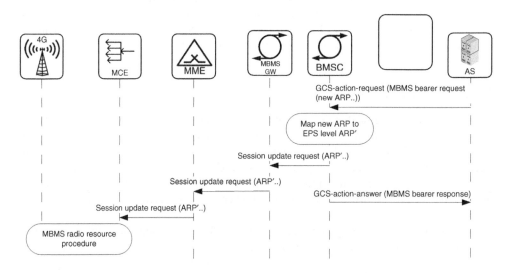

Figure 5.8   ARP modification request from GCS AS

because 3GPP TS 36.300 [10] requires that the Suspension/Resumption of MBMS service provisioning is applied to a whole MBSFN area.

If an MBMS Delivery is being suspended by the MCE, no notification is generated from BM-SC toward GCS AS. The GCS AS could only become aware of such an event by following Unicast Delivery requests from UEs due to the UE triggered service continuity procedure.

5.6.5 MBMS Delivery Status Notification

In Release-12, 3GPP has defined a generic procedure to allow sending of notifications from the BM-SC to the GCS AS related to the MBMS Delivery status. Currently, the BM-SC can indicate that MBMS Delivery was terminated due to expiry of the TMGI. It is expected that in future releases, this notification may be extended to include other possible information such as MBMS Delivery suspend notifications.

5.7 Establishment of MBMS Delivery

5.7.1 Pre-establishment

In this usage scenario, the MBMS bearer for MBMS Delivery is setup in certain areas before there is an actual need for it. For example, a parade is being scheduled in Seattle downtown to celebrate the win of the football season and it is expected that more than 500,000 fans will gather within the 10 miles parade route. The PS agency will mobilize hundreds of police officers and other emergency personnel for this event in order to ensure the required safety of this festivity.

The PS agency plans to setup four communication groups as follows: "police," "fire," "medical," and "general." In this example, the GCS AS will request the BM-SC to establish only one MBMS bearer for MBMS Delivery. The BM-SC establishes the MBMS bearer, allocates a TMGI, and provides it to the GCS AS. The GCS AS delivers group communication data for these four groups within the same MBMS bearer by multiplexing.

Figure 5.9 shows how GCS AS preestablishes a session for MBMS Delivery.

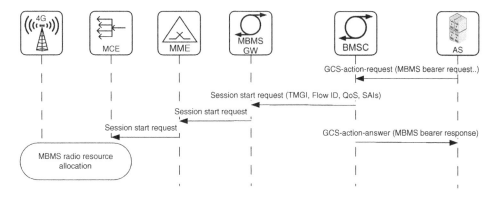

Figure 5.9 Preestablished session for MBMS Delivery

The GCS AS sends GCS-Action-Request (GAR) command to the BM-SC and includes the MBMS-Bearer-Request AVP, which contains the MBMS-StartStop-Indication AVP set to "START," the MBMS-Service-Area AVP, and the QoS-Information AVP. The "START" value indicates to the BM-SC that this is a request to start a new MBMS Delivery session. The MBMS-Service-Area AVP indicates a list of MBMS Service Area Identities (MBMS SAI). The MBMS service areas are preconfigured in the BM-SC and map to a list of serving cells. In this example, the MBMS SAIs are representing the cells that are serving the 10 miles parade route. The PS agency needs to gather information regarding service areas to be used from the PLMN operator. The QoS-Information AVP indicates the QCI value (only GBR QCI is allowed for MBMS bearers), the Guaranteed Bitrate for DL traffic, and the ARP indication. Because a single MBMS bearer is being used for the four groups, the GCS AS needs to determine the appropriate bitrate and ARP requirements to fulfill the needs of all four groups. The BM-SC uses these parameters to start the MBMS Delivery session and returns the allocated TMGI and Flow ID to the GCS AS in the MBMS-Bearer-Response AVP in addition to the session duration and user plane transport address. The GCS AS uses TMGI and Flow ID to identify the MBMS Delivery session. Session duration indicates the expiration time for this MBMS Delivery session. The GCS AS can request to renew the TMGI expiration time before expiry, if the MBMS Delivery has to be continued. The user-plane transport contains the IP address and port number in BMSC-Address AVP and BMSC-Port AVP where the GCS AS sends the media.

At this point the MBMS Delivery session is preestablished for the 10 miles parade route. The GCS AS multiplexes and encrypts the media for the four different groups to be delivered over the same MBMS bearer. The GCS AS also needs to configure each group member's UE with the relevant information to receive the group communication media for their respective group. These configuration data include the TMGI, decryption keys for the media, and the schema for demultiplexing the media in order to retrieve the correct group communication data sent over the same MBMS bearer (e.g., assigning different IP port number per group). This procedure is expected to be performed over GC1, which is not standardized in Release 12. Figure 5.10 summarizes the steps when a MBMS Delivery session is preestablished.

5.7.2 Dynamic Establishment

Dynamic establishment of an MBMS Delivery session is normally required due to unplanned incidents. For example, a bomb went off at a congested area and suddenly a large number of PS personnel (police, fire, and medical) are arriving to the scene and switching to group communication using MBMS Delivery.

Before the incident, the PS personnel are communicating with their peer group members via Unicast Delivery. The GCS AS has not activated the MBMS Delivery yet but has reserved a TMGI value for each group at the BM-SC. Figure 5.11 shows how the GCS AS reserves a TMGI for each group.

The GCS AS sends a GAR command to the BM-SC and the TMGI-Allocation-Request AVP indicates that it wishes to request three TMGI numbers. BM-SC allocates three new TMGIs, and a session duration indication that is common to all those three TMGIs. BM-SC provides these data to the GCS AS in the TMGI-Allocation-Response AVP of the GCS-Action-Answer (GAA) command. The GCS AS internally assigns one TMGI to each GCSE group, for example, TMGI-A=police, TMGI-B=fire, and TMGI-C=medical.

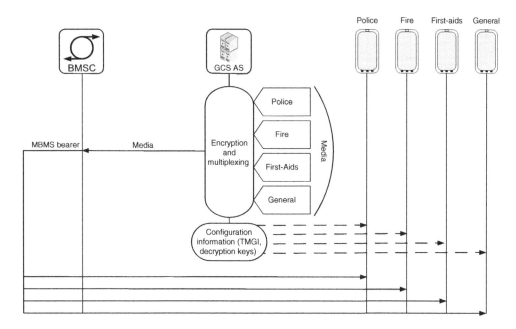

Figure 5.10   Steps for a preestablished session for MBMS Delivery

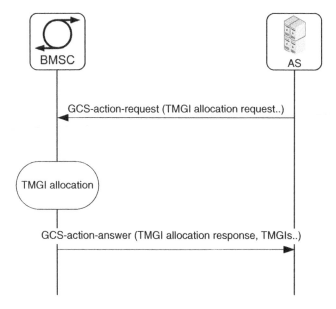

Figure 5.11   TMGI allocation

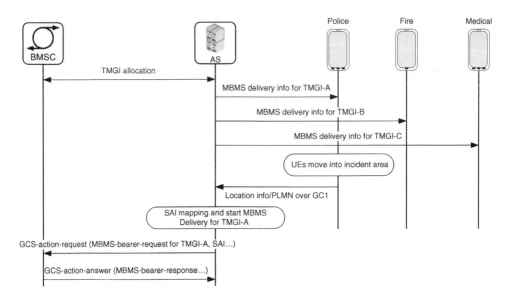

Figure 5.12   Activating MBMS Delivery session in a dynamic establishment scenario

The GCS AS provides MBMS Delivery configuration information to each of the member over GC1 interface. At this point, UEs will start monitoring their particular serving cell for the corresponding MBMS bearer for the MBMS Delivery while still continuing to maintain group communication using Unicast Delivery.

When the incident occurs and many group members (e.g., policemen) arrived at the scene, the GCS AS can decide that there are sufficient group members within the incident's area and to start MBMS Delivery. This decision can be based on information received via application layer signaling over GC1, where each group member is reporting his or her current location (e.g., geo-coordinates and/or cell identification and serving PLMN). To start MBMS Delivery in the incident's area, the GCS AS has to map the location area information to the MBMS Serving Area Identities configured for the serving PLMN. It is expected that this kind of mapping information is known beforehand, that is, as part of the service agreements between the GCS AS service provider and the serving PLMN operator. Figure 5.12 shows an example of activating the MBMS Delivery in a dynamic establishment scenario.

The MBMS Delivery activation command is the same that is used in the preestablishment scenario with the GAR command sent from GCS AS to BM-SC with the same parameters plus the additional TMGI indication. The BM-SC uses these parameters to start the MBMS Delivery session, returns the same TMGI to the GCS AS, and allocates a Flow ID that is provided to the GCS AS in the MBMS-Bearer-Response AVP in addition to session duration and user-plane transport address. The GCS AS uses the TMGI and Flow ID to identify the particular MBMS Delivery session and transmits group communication media to the user-plane transport address. Figure 5.13 shows the user plane aspects within a dynamic establishment scenario.

Basically, the MBMS Delivery activation will trigger the service continuity behavior in the UE as the UE notices that TMGI-A is now active in its serving cell and it can switch to

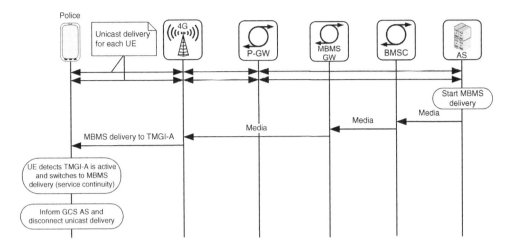

Figure 5.13   User plane for MBMS Delivery

MBMS Delivery to receive downlink media. As a result, the connection established for Unicast Delivery can be disconnected between each UE and the GCS AS, if it is not simultaneously used for uplink data transmission. This can help free up radio and network resources.

5.8   MBMS Delivery Procedures

5.8.1   MBMS Delivery Modification

When MBMS Delivery session has been established, either due to a preestablished or dynamic establishment scenario, the GCS AS may modify priority and preemption characteristics of the MBMS bearer and/or expand/reduce the service area where media are broadcasted.

One reason to modify the MBMS service area is that the incident (e.g., such as fire) may have spread to a wider area and thus expanding the MBMS service area would reduce the extra signaling for the UE to perform service continuity as the crossing of MBMS service area boundaries is minimized.

The priority and preemption characteristic defines how the MBMS bearer is treated at the time of congestion. For example, a MBMS bearer can be setup with low priority such that it can be preempted once a high-priority MBMS bearer is activated. One reason to modify the priority and preemption characteristic of an active MBMS bearer could be because the priority of the group using this bearer has changed; for example, that medic has now the highest priority because saving life is more important than catching the person who started the fire.

An existing MBMS bearer that has to be modified is identified by both TMGI and Flow ID. If a service area is to be expanded, the GCS AS must also ensure that within the new expanded area there is no ongoing MBMS Delivery session with the same TMGI. Otherwise, BM-SC will reject this type of modification because an MBMS service area cannot overlap with another MBMS service area with the same TMGI but different Flow IDs.

MBMS Delivery modification is triggered by GCS AS with a GAR command to the BM-SC. It includes the MBMS-Bearer-Request AVP, which contains a MBMS-StartStop-Indication

AVP set to "UPDATE" and the TMGI and MBMS-Flow-Identifier AVPs. The MBMS-Service-Area AVP is included, if the change of service area is requested. The QoS-Information AVP is included, if the priority and preemption characteristic is modified. Other QoS parameters (i.e., QCI, bitrate) remain unchanged, that is, the same values as during MBMS bearer activation are provided.

The BM-SC uses these parameters to start the MBMS Session Update procedure as defined in 3GPP TS 23.246 [3]. When the MBMS session is successfully updated, the BM-SC returns the same TMGI and Flow ID to the GCS AS in the MBMS-Bearer-Response AVP.

Although ARP modification is allowed via the MBMS Session Update procedure (see 3GPP TS 23.468 [14]), the bit rate information has to remain the same as the original value used during MBMS bearer activation. This restriction may prevent some group communication scenarios using MBMS Delivery, for example, when the media format changes from voice to voice plus high-definition video. To avoid switching from MBMS Delivery to Unicast Delivery and back to MBMS Delivery with higher bandwidth for voice plus HD video, 3GPP TS 23.246 [3] allows the GCS AS to activate a new MBMS bearer to replace the old one. This means the GCS AS can first initiate a second MBMS Delivery session with a new TMGI value allowing to transmit voice plus high-definition video and only after that inform the UEs via GC1 signaling to switch the group communication to the new TMGI. Afterwards the GCS AS can deactivate the old MBMS Delivery session with the old TMGI. To ensure service continuity, the UE must be able to simultaneously process multiple TMGIs and switch the downlink data content from the old to the new MBMS Delivery session. Figure 5.14 illustrates this scenario.

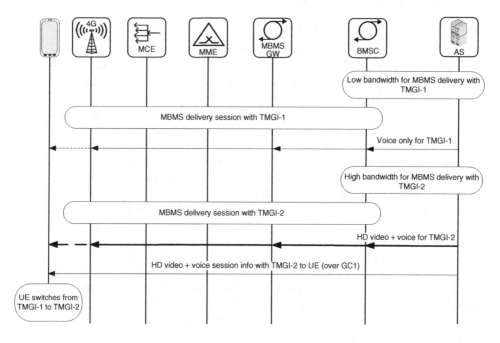

Figure 5.14   Switch between two MBMS Delivery sessions

5.8.2 MBMS Delivery Deactivation

MBMS Delivery session can be deactivated due to an explicit MBMS deactivation request from the GCS AS, explicit TMGI deallocation request from the GCS AS, local TMGI expiration at the BM-SC or triggered by Operation, Administration, and Maintenance (OA&M).

The GCS AS can deactivate MBMS Delivery by sending an explicit request to the BM-SC. This procedure is useful when an incident is over or a large number of group members have moved away from the incident area where the MBMS Delivery was previously activated. Similar to the dynamic establishment scenario, the GCS AS should be aware of how many UEs are being served by the MBMS Delivery session within the incident area. Deactivating MBMS Delivery arbitrarily may have unwanted side effects because those UEs, which are relying on MBMS Delivery, will switch to Unicast Delivery using the service continuity procedure. This could lead to a signaling storm within the serving network, which is highly undesired.

Figure 5.15 shows how the GCS AS can request the deactivation of MBMS Delivery.

The GCS AS sends a GAR command to the BM-SC and includes the MBMS-Bearer-Request AVP, which contains a MBMS-StartStop-Indication AVP set to "STOP" and the TMGI and Flow-Identifier AVPs of the MBMS Delivery to be deactivated.

BM-SC deactivates the MBMS bearer using the MBMS Session Stop procedure as specified in 3GPP TS 23.246 [3]. The BM-SC acknowledges the request by returning the same TMGI and Flow ID as received in the request from GCS AS in the MBMS-Bearer-Response AVP within the GAA command.

If MBMS Delivery is deactivated due to TMGI expiry, the BM-SC will initiate a MBMS Session Stop procedure to release the MBMS resources. In this case, the BM-SC will notify the GCS AS that the MBMS Delivery (identified by TMGI and Flow ID) has been deactivated. Figure 5.16 shows this notification procedure.

When MBMS Delivery is deactivated due to TMGI expiry, the BM-SC sends a GCS-Notification-Request (GNR) command including MBMS-Bearer-Event-Notification AVP with TMGI and Flow-Identifier AVPs to indicate the MBMS Delivery session that has been deactivated. The GCS AS acknowledges with a GCS-Notification-Answer (GNA) command. There is no explicit notification sent from BM-SC to GCS AS in the TMGI expiry scenario if there is no corresponding MBMS bearer being activated. If MBMS Delivery is deactivated due to explicit TMGI deallocation requested by the GCS AS, the BM-SC will initiate the MBMS Session Stop procedure for that particular TMGI.

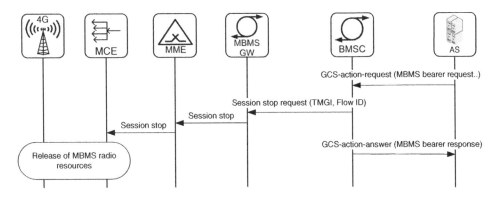

Figure 5.15 MBMS Delivery release due to GCS AS request

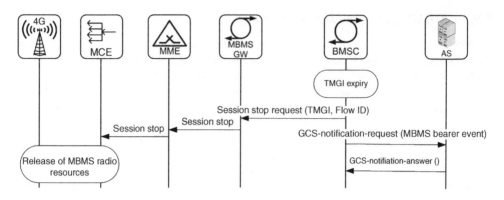

Figure 5.16 MBMS Delivery release due to TMGI expiry

If MBMS Delivery is deactivated due to explicit TMGI deallocation requested by the GCS AS, the BM-SC will initiate the MBMS Session Stop procedure for the particular TMGI.

If MBMS Delivery is deactivated triggered by OA&M, the 3GPP specification does not mandate an explicit notification message from BM-SC to GCS AS. However, it is expected that the BM-SC treats this event like the TMGI expiry scenario in order to make the GCS AS aware of the session deactivation.

5.8.3 TMGI Management

For both the preestablished and dynamic establishment MBMS Delivery scenarios, the TMGIs are allocated by the BM-SC and are always tied to an expiration time. When a TMGI expires,

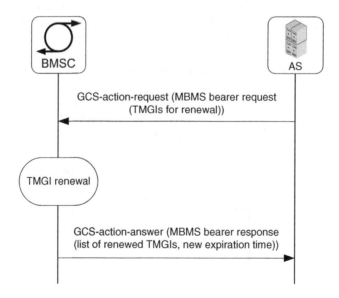

Figure 5.17 TMGI expiry timer renewal

MBMS resources associated with that TMGI (regardless of the Flow ID) would be removed by the BM-SC using the MBMS Session Stop procedure as specified in 3GPP TS 23.246 [3].

The GCS AS can request to renew the TMGI with the procedure shown in Figure 5.17.

The GCS AS sends a GAR command to the BM-SC. The included TMGI-Allocation-Request AVP indicates the TMGIs for which the expiration times have to be renewed. If the number of TMGIs being requested for renewal is zero, all TMGIs that are currently allocated to the GCS AS should be renewed.

The BM-SC indicates a new expiration time and the list of TMGIs being renewed back to the GCS AS in the TMGI-Allocation-Response AVP of the GAA command. The provided expiration time is common to all renewed TMGIs.

The GCS AS can also explicitly indicate to the BM-SC that the already allocated TMGIs are no longer needed by using the TMGI-Deallocation-Request AVP in the GAR command. If any of those TMGIs is still active (i.e., MBMS Delivery with the TMGI is ongoing), the BM-SC will deactivate the corresponding MBMS bearer implicitly. However, an explicit instruction (i.e., deactivate MBMS Delivery before deallocating the corresponding TMGI) can be used to ensure that the network is being properly designed and deployed while the implicit deactivation procedure can be used to avoid hanging of resources. Therefore, it is recommended that the GCS AS always deactivates the MBMS Delivery session before requesting it from being released by the BM-SC.

5.9 Access Control

Priority and preemption are essential functions in PS LTE networks. However, these functions can only be applied by the network in case the UE is already able to send a connection request over the radio interface. In the event of a major incident, many users may try to access the LTE network at the same time. This can create congestion at the radio access level, with the consequence that some UEs may be rejected due to lack of resources and in the worst case some may not even get a response from the network. Those UEs that were not able to access the network will have to retry. With the concept of Access Classes (ACs), UEs are divided into different classes allowing the operator to restrict access to the (radio) network in high-load or congestion situations to UEs belonging to certain classes. Access Class values AC0 to AC15 are defined and stored on the Universal Subscriber Identity Module (USIM). While AC0 to AC9 are randomly assigned ordinary classes, AC10 to AC15 are used for special purposes (details are explained further below).

An AC determines the request indicator that can be used by the UE for gaining access to the LTE radio network. More specifically, the AC value of the UE is an important criteria to determine the RRC establishment cause that can be used within the RRC Connection Request sent to the eNB for establishment of a radio connection (see 3GPP TS 24.301 [15] and 3GPP TS 36.331 [11]). If the UE belongs to AC11 to AC15, it can provide "HighPriorityAccess" as establishment cause in the RRC Connection Request. The eNB uses the RRC establishment cause to prioritize a connection request from such UE(s). This implies that requests with "High-priority access AC11 – AC15" indication from high-priority users will be treated by the network with higher priority than requests from regular users.

Furthermore, when the network is in a high congested situation, the operator may have to activate Access Class Barring (ACB) in one or more eNBs to protect the network from extreme failure or shutdown and at the same time ensure those "high-priority" UE(s) will be able to gain access to radio resources for network access. ACB can be invoked on a cell-by-cell basis. Whether a certain UE is barred for radio access or not is determined by the access class of the UE.

Therefore, it is important for PS networks to assign the access class(es) to PS users in a manner allowing "high-priority" UE(s) to gain access to the network even in high-load situations. Access to the network is, for example, essential when the PS user is switching from MBMS Delivery to Unicast Delivery to ensure service continuity. Access control becomes more important also in case the network is used by both PS and non-PS personnel.

However, Access Class assignment to PS users is not as simple as assigning access classes to commercial users. In a normal commercial LTE network, a user is assigned an Access Class between 0 and 9 randomly. If a user is also part of a specific high-priority group such as PLMN staff, public utilities, and so on, then they will be assigned an additional Access Class between 11 and 15. Hence, the relationship between normal user and high-priority user in a commercial LTE network is to some extent static. In PS networks, this may not be the case. PS networks are expected to be shared by many different types of agencies (e.g., police, fire, medic, public utility) and each particular incident may dictate a certain agency to be the most critical PS user (e.g., police is given high priority for parade arrangements, fire for a forest fire, and medic for handling medical emergency situations). In other words, PS operation and incidents are very dynamic and assignment of a static single Access Class to a user may not meet all the operational needs at any time. Therefore, it is important to understand how ACB works in LTE networks allowing certain users to access the LTE network even in extreme congestion situations. Thus, we provide in the following a description of this feature.

All UEs are member of one out of 10 randomly allocated Access Classes between 0 and 9, which are preconfigured on the USIM. In addition, the UE can be assigned one or more of the five special Access Classes between 11 and 15. In 3GPP TS 22.011 [16], the special Access Classes are defined as follows:

- Access Class 15 – PLMN staff
- Access Class 14 – Emergency Services (e.g., police, firemen, or medic)
- Access Class 13 – Public Utilities (e.g., water and gas suppliers)
- Access Class 12 – Security Services
- Access Class 11 – For PLMN Use

The PLMN operator can (re-)assign any of these Access Classes on the USIM via the Over-the-Air (OTA) provisioning method, which is based on SMS as transport technology. AC11 and AC15 are only valid at the HPLMN, while AC12 to AC14 are valid at the HPLMN and VPLMNs of the home country. The UE is located in the home country when the Mobile Country Code (MCC) of the IMSI matches the MCC of the serving PLMN. A country is normally identified by a single MCC value, with the exception that MCC values 310 through 316 identify a single country (United States) and MCC values 404 through 406 identify a single country (India). AC10 is a special class to control emergency calls and not assigned to the USIM. AC10 is signaled over the air to indicate whether or not network access for emergency calls is allowed for UEs with access classes 0 to 9. For UEs with access classes 11 to 15, emergency calls are not allowed if both "Access Class 10" and the relevant Access Class (11 to 15) of the UE are barred. Otherwise, emergency calls are allowed for these UEs.

Access Class Barring can be invoked individually for AC11 to AC15. In other words, the network can allow UEs with AC14 to access the network, while barring access of UEs with AC11, AC12, AC13, and AC15. However, ACB cannot be controlled individually for AC0 to AC9. All UE(s) with Access Classes between 0 and 9 are subject to ACB when it is activated.

In order to enable further granularity in activation of ACB and reduce disruption of services, 3GPP introduced a feature called Smart Congestion Mitigation (SCM) for LTE in Release 12. With this feature, the network can indicate whether a specific application or service is exempted from ACB (commonly referred to as "ACB skip"). In other words, the network can broadcast an "ACB skip indicator" for certain applications or services (e.g., multimedia telephony voice or video or SMS) and allow access for such applications or services while other applications and services are barred due to ACB. So, when ACB is activated, the UE should check if the application or service is allowed to access the network based on the "ACB skip indicator" broadcasted by the network (for details see 3GPP TS 24.301 [15] and 3GPP TS 36.331 [11]). Based on that, the UE can determine whether it can send a request to establish a radio connection toward the eNB.

5.10   Mission Critical Push To Talk

MCPTT is a service, which requires a specific application in the UE and network. It mimics the behavior of the PTT services provided by legacy LMR (LMR) systems. A PTT application provides an arbitrated method by which two or more users can engage in communication. Users have to request the permission to transmit, traditionally by pressing a button on their handset. Only one user can talk at a certain time and all others are just listeners.

Considering the likely deployment scenarios of PS communication based on LTE it is obvious that in many scenarios the LTE-based PS network will run in parallel to legacy PS networks (e.g., a P.25 or TETRA based system), thus the PS UE will have to be able to support both, LTE and legacy techniques and first responders will be using both depending on, for example, coverage or load conditions. For ease of use, it is expected the UE will change between the systems automatically and the user is not being aware of whether LTE or legacy network is currently used. In this case it is very important that the MCPTT service behavior closely resembles the well-known legacy service behavior in order not to confuse users by different behavior when operating in one or the other network.

To allow for this, both the P.25 and the TETRA communities are participating actively in 3GPP work on MCPTT service definition and provide their views and concrete requirements to 3GPP in order to achieve seamless service behavior for users when operating under different PS systems.

The service requirements are collected in 3GPP TS 22.179 [17] within 3GPP Release 13. Work on this stage 1 technical specification is expected to conclude by end of 2014, while stage 2 and 3 work is only about to start at the time of writing this book, but it is expected to conclude within the Release 13 timeframe. Enhancements will likely be specified in future releases.

Normally, 3GPP has left layer specification work in Release 12 to other groups or even to individual company implementations. As there is a strong desire by the PS community to have a globally unified and standardized MCPTT service architecture in place, 3GPP in Release 13 has decided to enter the service layer area for once. It is obvious that it will take some time for the different views from P.25 and TETRA communities to converge on their requirements first. It is foreseeable that MCPTT service will have to support many options and requirements, thus leading to a complex system.

The MCPTT service (or just MCPTT) makes use of the previously described enablers GCSE and ProSe in parallel to other functionality provided by EPS such as priority and preemption

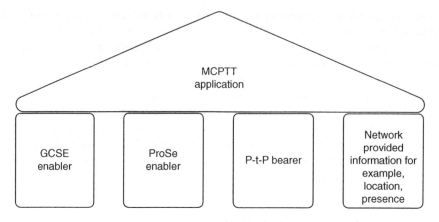

Figure 5.18 Relation of MCPTT to ProSe and GCSE

mechanisms, location services, and the like. Figure 5.18 provides an overview about the relation between MCPTT and GCSE, ProSe.

MCPTT will operate on devices with E-UTRAN network coverage, but also on devices making use of ProSe communication without the involvement of any network infrastructure or on devices working as ProSe relays for other MCPTT-enabled devices. The end user's experience of the service should be the same regardless if MCPTT is used on a device under network coverage or based on ProSe without network coverage.

The MCPTT can also be used by non-PS users such as employees of utility or railway companies.

5.10.1   MCPTT Service Description

This section is based on excerpts from 3GPP TS 22.179 [17]. It describes the most important requirements for MCPTT.

The MCPTT is intended to support communication between several users (a group call), where each user has the ability to gain the right to talk in an arbitrated manner. MCPTT will also support "private calls" between pairs of users. The MCPTT is expected to be built on top of the techniques provided by LTE/SAE to establish, maintain, and terminate the actual communication path(s) (i.e., the user plane) among the users.

MCPTT allows users to request the "right to talk" and provides a mechanism to arbitrate between requests that are in contention. This is called "floor control." In case of multiple requests the determination of which user's request is accepted and which users' requests are rejected or queued is based on a number of characteristics such as their priorities. MCPTT provides a means for a user with high priority (e.g., in an emergency condition) to interrupt the current talker. MCPTT also supports a mechanism to limit the time a user is able to talk thus permitting users of the same or lower priority a chance to gain the floor.

The MCPTT provides means for a user to monitor several calls in parallel and enables the user to switch focus to a selected call. A user may join an already established MCPTT group call; this is known as "late call entry." In addition, MCPTT provides identification of the current talker to the listeners and location determination to users.

Members of a PS network are organized in groups. People that are working together communicate in the same MCPTT group, the group communication helping them to coordinate quickly.

People with different tasks often communicate in separate groups. Communication structures and MCPTT groups are prepared for the handling of large incidents and control of large events. Thus, there need to be MCPTT groups and procedures for coordination with PS personnel from other organizations and/or even other countries. This results in a large number (e.g., more than 100) of MCPTT groups available to PS personnel, from these groups the PS user can select which groups she/he wants to affiliate ("Affiliated Groups") with at the moment, that is, to receive communication from. Furthermore, the PS user can select which group his communication will be directed to. This is the so-called Selected Group.

5.10.2   MCPTT Call Types

The application for MCPTT has to support a number of "special" group calls including: "Broadcast Group Call", "Emergency Group Call,", and "Imminent Peril Group Call".

For Broadcast Group Calls, broadcast does not refer to the distribution mechanism used on the radio but to the fact that the initiating MCPTT user expects no response from the receiving MCPTT users. Consequently, receiving users cannot respond to a Broadcast Group Call. The intended receiving MCPTT users may be a subset of all MCPTT users or all users in a system. A System Call is a special case of a Broadcast Group Call that is transmitted to all users in a dynamically defined geographic area.

Emergency Group Calls are initiated by users that are in a, for example, life threatening, emergency situation to call for help and have the highest priority over all other MCPTT group transmissions, except for System Calls and other MCPTT Emergency Group Calls.

Imminent Peril Group Calls are calls prioritized in the event of an immediate threat, for example, forest fire about to encircle campers, tanker truck ready to explode near a school, casualties at scene of a car bombing. Imminent Peril Group Calls have priority over all other MCPTT Group transmissions, except System Calls, Emergency Group Calls, and other Imminent Peril Group Calls.

Private Calls are two-way calls between a pair of MCPTT users using the MCPTT service together with floor control. Private Calls leverage many of the features of MCPTT Group Calls, such as providing of MCPTT user identity and alias information, location information, encryption, privacy, priority, and administrative control.

5.10.3   MCPTT Priorities

The priority and QoS model of 3GPP falls short when it comes to fulfilling the needs of MCPTT.

MCPTT Priority and QoS are situational. The MCPTT is intended to provide a real-time priority and QoS experience for MCPTT calls, as PS users have significant dynamic operational conditions that must be accounted for when determining their priority. For example, the type of incident a responder is serving or their overall shift role strongly influences a user's ability to obtain resources from the LTE system.

The MCPTT priority model identifies three areas of prioritization: prioritization between and within sessions, intersystem prioritization, and prioritization within the network (between EPS and UE). At the application layer, a functional entity called "MCPTT Priority and QoS Control" processes with each request static preconfigured information about users and groups participating in MCPTT as well as dynamic (or situational) information in order to set dynamically the appropriate priority and QoS for a MCPTT call.

The "User Static Attributes" include information categorizing the user by several criteria including whether the user is a first or second responder, supervisor, dispatcher, or administrator and a preconfigured system-wide individual priority level for that user.

The "Group Static Attributes" include information about the nature and type of the group and the owning agency, such as the minimum priority characteristics that are guaranteed to all the participants in communication associated with this group, regardless of their individual priority characteristics.

The "User Dynamic Attributes" include, among others, the user's operational status, the type of incident (e.g., MCPTT Emergency or Imminent Peril) the user might be involved in, and a number of other parameters that may change at any time during or in between calls.

The "Group Dynamic Attributes" include the type of incident (e.g., MCPTT Emergency or Imminent Peril), if any, the group is currently involved in and a number of other parameters that may change at any time.

5.10.4   Shareable MCPTT Devices

Up to now the basic working assumption in 3GPP was a one-to-one relationship between a UE and a user. This model needs to be extended for shareable MCPTT UEs to allow a pool of UEs, each UE being interchangeable with any other, and users randomly choosing one or more UEs from the pool for temporary exclusive use. A shareable MCPTT UE can be used by anyone gaining access to the MCPTT application on the UE and thus, becoming an authenticated MCPTT User. A shareable MCPTT UE can serve only one MCPTT user at a time. An MCPTT user who signs into a shareable MCPTT UE that is already in-use causes the sign-off of the other MCPTT user.

An MCPTT user can simultaneously be assigned to several active MCPTT UEs, which, from an MCPTT service point of view, are addressable individually and/or collectively within the context of their association to the MCPTT user. In other words, the responder identity is not tied to the USIM. For example, each of the responders in a group will receive an LTE capable UE and it is expected that each user will then login at the application, for example, with their unique user-id and password. After successful login, the UE will run their customized applications with the according priority level.

5.10.5   On and Off Network Mode of Operation

As mentioned earlier, MCPTT is able to operate in two modes. The first one is the "on network operation mode" when MCPTT utilizes the EPS and specifically, but not limited to, the GCSE service enabler. The second mode of operation is the "off-network mode" using a ProSe direct (UE-to-UE) communication path. The ProSe direct communication path does not traverse the network infrastructure.

Consequently, MCPTT requirements are separated to be applicable in case of on-network operation, in case of off network operation, and requirements that apply to both modes of operation.

5.10.6   *Interworking with legacy PTT Systems*

Mission critical users currently employ a wide range of narrowband mission critical push-to-talk services. Project 25 (governed by the TIA-102 standards) and TETRA (governed by ETSI standards) are digital PS PTT systems. In addition, "legacy" or "conventional analog" systems are common throughout the world. These systems provide PTT and related services that are analogous to those provided by MCPTT, including group calls, private calls, broadcast calls, dynamic group management, and other services.

MCPTT is intended to interwork with these non-MCPTT systems.

References

[1] 3GPP TS 22.468: "Group Communication System Enablers for LTE (GCSE_LTE); Stage 1".
[2] 3GPP TS 33.246: "Security of Multimedia Broadcast/Multicast Service (MBMS)".
[3] 3GPP TS 23.246: "Multimedia Broadcast/Multicast Service (MBMS); Architecture and Functional Description".
[4] 3GPP TS 32.273: "Charging management; Multimedia Broadcast and Multicast Service (MBMS) charging".
[5] 3GPP TS 23.203: "Policy and charging control Architecture".
[6] 3GPP TS 29.468: "Group Communication System Enablers for LTE (GCSE_LTE); MB2 Reference Point".
[7] RFC 3588: "Diameter Base Protocol".
[8] Void.
[9] 3GPP TS 29.061: "Interworking between the Public Land Mobile Network (PLMN) Supporting Packet Based Services and Packet Data Networks (PDN)".
[10] 3GPP TS 36.300: "Evolved Universal Terrestrial Radio Access (E-UTRA) and Evolved Universal Terrestrial Radio Access (E-UTRAN); Overall Description".
[11] 3GPP TS 36.331: "Evolved Universal Terrestrial Radio Access (E-UTRA); Radio Resource Control (RRC)".
[12] 3GPP TS 36.304: "Evolved Universal Terrestrial Radio Access (E-UTRA); User Equipment (UE) Procedures in Idle Mode".
[13] 3GPP TR 36.868: "Study on Group Communication for E-UTRA".
[14] 3GPP TS 23.468: "Group Communication System Enablers for LTE (GCSE_LTE); Stage 2".
[15] 3GPP TS 24.301: "Non-Access-Stratum (NAS) protocol for Evolved Packet System (EPS); Stage 3".
[16] 3GPP TS 22.011: "Service accessibility".
[17] 3GPP TS 22.179: "Mission Critical Push to Talk MCPTT (Release 13)".

6

Summary and Outlook

6.1 Role of LTE

Nowadays Long-Term Evolution (LTE) is the worldwide de facto standard for mobile broadband communication and it is expected to be in this position for the next decade, although research work is ongoing in the industry for a new radio technology (working title 5G). For commercial network operators, spectrum flexibility and efficiency, Operational Expenditure (OPEX) and Capital Expenditure (CAPEX) reductions due to a simplified architecture, and consequent usage of the Internet Protocol suite throughout all network interfaces are key points (i.e., there is no need to have a team of operational experts for Internet Protocol (IP) and non-IP protocols any more, for example, ITU-T signaling system No. 7 based protocols). Public Safety networks adopting LTE will benefit from these advantages. They can deploy a mobile system that has proven its reliability, flexibility, security, and feature-richness in mass deployments (forecasts predict 2.5 Billion LTE subscriptions by end of 2020). As the de facto mobile broadband standard in all parts of the world, it provides the necessary economy of scale in terms of shipped chipsets, devices, and network equipments. Backed by a strong standardization organization like 3rd Generation Partnership Project (3GPP), necessary changes to the LTE system are provided rapidly and based on consensus among all major device and infrastructure vendors. The development of LTE is far from an end. 3GPP is constantly improving and optimizing the whole system. New features are added with each release, new frequencies are added to the LTE bands, and bandwidth is even increasing with LTE Advanced (LTE-A), Carrier Aggregation (CA), and Multiple-Input-Multiple-Output (MIMO). As a matter of fact, 3GPP's efforts are very much focused on improving LTE, not so much anymore on improving older technologies such as Universal Mobile Telecommunications System (UMTS) or General Packet Radio Service (GPRS). This provides confidence to network operators that necessary resources in 3GPP are available to correct errors, optimize, and evolve LTE. Optimizing LTE for Public Safety in Release 12 has proven 3GPP's capability and willingness to take new requirements on board and optimize or adapt LTE also for new use cases. Other examples are optimizations to fit LTE for new market opportunities around Machine-to-Machine (M2M) communication and the Internet-of-Things (IoT). Future enhancements of LTE (more specifically to the Evolved Packet Core) due to the introduction of Network Function Virtualization (NFV) and Software Defined Networking (SDN) techniques will allow Public Safety network

LTE for Public Safety, First Edition. Rainer Liebhart, Devaki Chandramouli, Curt Wong and Jürgen Merkel.
© 2015 John Wiley & Sons, Ltd. Published 2015 by John Wiley & Sons, Ltd.

operators to benefit from increased deployment flexibility, reduced CAPEX, and OPEX due to ease of scaling up and down network resources and capabilities. Other features 3GPP is currently working on in Release 13 like paging optimizations will help Public Safety networks to prioritize voice calls, thereby reduce call setup times and help to meet their strict latency requirements. Such feature enhancements will come "for free" for Public Safety networks due to the fact that LTE is a system that is specified by a global standardization organization.

Network sharing and (national/international) roaming will allow Public Safety operators to deploy dedicated networks only in some parts of a country but have sharing or roaming agreements with commercial LTE operators to use their network (especially Radio Access) infrastructure. This can lead to significant cost reductions while building up a nationwide Public Safety network. On the other hand, commercial operators will benefit from providing their network capacities, infrastructure, and know-how to the Public Safety community.

The end user is mainly interested in getting fast and reliable access to IP-based services such as web browsing, voice and video calls, video streaming, online gaming, and much more. LTE provides access to this kind of services with high bandwidth and low latency, combined with the benefits of a standardized cellular system, including nomadicity, mobility, and worldwide roaming agreements. For many people, LTE in combination with innovative new tariff structures has brought the vision of being "always-on" (e.g., connected to the Internet at any time) to reality. Once LTE is rolled out in more countries this vision will become reality for much more people in the future.

6.2   Public Safety Features

As explained throughout this book, LTE provides a big variety of capabilities enhancing people's safety in their daily life. Although the focus of this book was on LTE for "Public Safety networks" we provided also insights about other public safety-related features that are available in LTE networks. One of the most important ones for the regular public is certainly the emergency call. Especially when provided in future via 3GPP's IP Multimedia Subsystem (IMS), it allows people not only to setup an emergency voice call but also using other media such as video call, messages or voice, and video clips to communicate with emergency call centers. In combination with enhanced location services (e.g., Location-Based Services, Assisted Global Positioning System (GPS)), this new kind of multimedia emergency services will allow for more accurate and faster assistance for people in emergency situations. Even before rescue personnel has reached the location where an accident or a disaster has occurred, pictures and videos were sent via a multimedia emergency service to the emergency service center and eventually to the rescue personnel on their way to the accident and will help to prepare them with more precise information, for example, what situation they can expect when arriving at the accident or disaster location. This can save the lives of people involved in an accident but also the lives of the rescue personnel as they have a clearer and more accurate view of the situation in advance.

The eCall system (see Ref. [1] for more information) that will be introduced in the European Union provides information to the emergency center automatically from cars that are involved in an accident. Besides providing data such as car's position, number of passengers, and speed before the accident, eCall can call rescue personnel directly even when the passengers of the car are not able to communicate anymore. Once eCall is available via LTE and IMS much more data at higher speeds can be sent to the emergency center or to a data center. Cars that may be equipped in the future with lot of sensors and cameras can send precise pictures or real-time videos to the emergency service center or to rescue personnel directly. Similar features like eCall for cars can be also foreseen for buildings or certain confined places. Once a fire is detected in a house, for example, in a public building, not only a simple fire alarm but also pictures or a real-time video stream showing the fire and the rooms where it is ongoing can be sent. This will help firemen to coordinate their operation before and after arrival at the accident site.

Public Warning System (PWS) allows an official authority to send warning messages over LTE to people in a certain area where a disaster (tsunami, earthquake, or volcanic eruption) may occur in the near future. Again, public warning messages in combination with a mobile broadband system such as LTE are a much more powerful method to make people aware of the dangers that are ahead of them, and more precise data regarding the area and degree of the disaster can be delivered to the endangered people.

Looking ahead we can even assume new Smartphone Apps are available, providing safety-related information to users: information coming directly from authorities where the App is registered but also information from other users who may be near a disaster or accident and inform either everyone or only friends in real time about the situation. This could lead to a setup of new kinds of social networks: social networks providing participants with safety relevant information wherever they are.

6.3 LTE for Public Safety

Proximity Services (ProSe) and Group Communication System Enablers (GCSE) are the two remarkable features standardized by 3GPP to allow Public Safety communication over LTE. With proximity services, fire and policemen and other Public Safety personnel are able to detect and communicate with members of a group they are belonging to without network coverage. The principle of device-to-device communication using LTE frequencies is a new kind of feature to 3GPP. Commercial network operators offer their services usually or only via their network infrastructure using licensed spectrum allocated to them by national authorities. This type of walkie-talkie function (i.e., 1-1 or 1-many communication) is essential for public safety personnel in a disaster region where none of the devices or only some of them are in network coverage. Devices in coverage can act as communication relays for devices out of coverage. The group communication enabler based on Multimedia Broadcast/Multicast Service (MBMS) as described in Chapters 1 and 5 is important for providing a resource efficient way for "Push-to-Talk" (PTT) communication where only one person can speak at a given time to all other members of his or her group. Some improvements of the MBMS feature were necessary to allow small talk bursts to be delivered to a certain area (a probably small area like a single cell) in a fast and efficient way. Fast and efficient communication refers to communication with low ("mouth to ear") latency and without wasting network resources so that many different groups, for example, groups of fire and policemen, can use the network in a distinct area while communication of other people is still possible.

An important aspect for Public Safety is the reliability of the underlying network and applications on top. LTE provides in-built reliability and robustness, for example, through restoration features on a per node (Mobility Management Entity (MME), Serving Gateway (S-GW), Packet Data Network Gateway (P-GW), Policy and Charging Rules Function (PCRF), Multimedia Broadcast/Multicast Service Gateway (MBMS GW)) level. Restoration features provide capabilities on protocol level to restore certain context data and sessions in a node such as the MME, S-GW, or MBMS GW after restart of this node.

The PTT application as such is not part of 3GPP's Release 12 activities. However, PTT is an integral part of existing Public Safety systems such as P25 (Project 25, a set of digital radio standards for use by Public Safety agencies in North America) and TETRA (Terrestrial Trunked Radio, a time division multiplex system standardized by ETSI), which are narrowband mobile radio systems. As a consequence, there is a strong desire to develop the PTT application on top of the group communication system enabler specified in Release 12.

The application known as Mission Critical Push-To-Talk (MCPTT) is currently being worked on in 3GPP Release 13. Completion of this work can be expected in late 2015. Service requirements for MCPTT are listed in 3GPP TS 22.179 [2]. These requirements consist of a general description of the PTT feature including floor control, group call, announcement of group call, group management, and priority of group calls but also requirements related to the allowed latency of group calls and the possibility to interwork with existing systems such as P25 and TETRA. Strict latency requirements for the MCPTT service will require using special bearer Quality of Service (QoS) for MCPTT signaling and user plane fulfilling strict QoS (low packet delays) and priority requirements. At the time of writing this book, it seems quite likely that MCPTT will be based on the Push-To-Talk-Over-Cellular (PoC) application specified by the Open Mobile Alliance (OMA) with several enhancements and optimizations to allow for mission critical communication in the field of Public Safety. Obviously, the MCPTT application has to run on top of the Group Communication System Enabler as specified in 3GPP Release 12 and has to implement the respective interfaces, that is, the MCPTT application server has to implement the 3GPP-specified interface toward the BM-SC (MB2 interface for control and user plane). OMA PoC control plane is based on SIP and it can particularly rely on IMS capabilities (e.g., with regard to reuse IMS security, registration, session establishment), which would make it easier to integrate an IMS-based MCPTT service with VoLTE on the same device and chipset. OMA PoC specifies how a PoC client registers for the service with a PoC server, initiates a PoC (on-demand or pre-established) session, joins or leaves a session, and terminates a session. The OMA PoC user plane specifications define, for example, the media/talk burst protocols allowing a PoC client to request the right to send talk and multimedia bursts to other clients. The latest versions of OMA PoC enabler specifications version 2.1 can be found in Ref. [3].

In addition, MCPTT need to make use of ProSe capabilities when device-to-device or device-to-network relay communication is required. Several possible solutions for the relay of MCPTT communication from device to device are discussed: either a Layer 2 (i.e., relay on the link layer) or Layer 3 relay (i.e., relay on the IP layer, this can, for example, make use of the well-known IPv6 prefix delegation feature), an application layer gateway (i.e., a MCPTT gateway/proxy function) running in the relay device, or a relay of the multicast transport (i.e., the relay device acts like a base station forwarding data received on a multicast channel to surrounding devices on another multicast channel). Discussions on possible solutions have just started at the time of writing this book.

3GPP has also started work on another feature called "Isolated E-UTRAN Operation for Public Safety (IOPS)." IOPS use cases are listed in 3GPP TR 22.897 [4]. The requirements for IOPS can be found in 3GPP TS 22.346 [5]. At the time of writing this book, work on stage 2 (architecture work) and stage 3 (protocol work) has not commenced yet so the description about IOPS is subject to future changes.

IOPS aims at providing services to Public Safety users even when the network is not fully functional, for example, when eNodeBs have become disconnected from other parts of the network and form an "Isolated E-UTRAN." Situations like that predominantly occur when there are natural disasters such as earthquakes or tsunamis. Under these conditions, ensuring the continued ability of Public Safety users for mission critical communication is of utmost importance.

IOPS also aims to provide the ability to deliberately create a local serving radio access net work without backhaul connections, by deploying one or more additional so-called Nomadic eNBs (NeNBs) working in a stand-alone mode. This is intended for use in areas where there is no or very limited coverage, for example, in case of a bush fire in a remote region. In this case, an Isolated E-UTRAN may comprise of one or more NeNBs. A NeNB should be able to initiate Isolated E-UTRAN operation under the control of the operator (details of how this can be realized are left to further work in 3GPP). An Isolated E-UTRAN derived from NeNBs is expected to exhibit similar behavior like an Isolated E-UTRAN derived from eNBs including: support for Public Safety UEs in the coverage area, communication between NeNBs, and support for limited backhaul connectivity.

In addition, the IOPS feature also aims to address a scenario where a fixed or nomadic set of eNBs is without normal backhaul communications but has been provided with an alternative (non-ideal) limited bandwidth backhaul, for example, only able to support signaling traffic but very little or no user plane traffic. Again, its purpose is mainly for use in areas where there is very limited or no coverage and there is, for example, only a satellite backhaul connection available for backhauling.

The Isolated E-UTRAN may comprise a single or multiple eNBs, a single or multiple NeNBs, or a mixed group of eNBs and NeNBs. This mixture of eNB and NeNB can occur when there is an outage due to a disaster and additional NeNBs are deployed for improving coverage and capacity in the Isolated E-UTRAN area. An Isolated E-UTRAN comprising multiple (N)eNBs, with connections between the (N)eNBs, can provide communication between UEs across a wider area of coverage that can be provided by a single isolated (N)eNB. The UEs in the coverage of the Isolated E-UTRAN are expected to continue communication and provide a restricted set of services supporting voice, data, and group communications to their Public Safety users. IOPS should be able to support Isolated E-UTRAN networks joining with each other to form larger, but still Isolated E-UTRANs, for example, when recovery is ongoing after a natural disaster and connections between eNodeBs are reestablished again step by step.

While under coverage of an Isolated E-UTRAN, MCPTT users should be offered some limited functionality to make use of the Isolated E-UTRAN for Public Safety communication. As there is no EPC available, group memberships and affiliation to these groups has to be pre-provisioned to take effect when operating in an Isolated E-UTRAN scenario. As no dynamic setup of groups is possible under this scenario, these groups will have to be fairly generic (e.g., some of them even including users from other organizations). For example, there may be a group defined including police, firefighters, and paramedic personnel, something that would usually be created on the spot and only if needed at an accident scene.

A Public Safety user should also be offered the means to initiate a MCPTT Emergency Group call that may reach a dispatcher (the dispatcher can be connected directly to the (N)eNB or via a UE), members of the group under the coverage of the Isolated E-UTRAN, and/or all UEs under the coverage of the Isolated E-UTRAN.

During Isolated E-UTRAN operation, each group can initiate at least one MCPTT voice group communication that all authorized members are allowed to join, up to the limit of available radio resources.

6.4 Outlook

We have seen that 3GPP has made remarkable efforts within only one release cycle to provide necessary features and enablers for the Public Safety community. These features are called Proximity Services, that is, device-to-device communication over LTE and Group Communication System Enablers. The latter one is based on MBMS for LTE. Owing to time constraints, not all aspects of these features could be developed within the Release 12 time frame. In the next 3GPP Release(s), an MCPTT application, enhancements, and optimizations for Proximity Services and Group Communication will follow. We expect first deployments of Public Safety networks in 2015, although trial networks in specific regions (e.g., in some US states or counties) may already be operated in 2014. The status of ongoing and planned 3GPP work items regarding LTE for Public Safety (authority-to-authority) communications can be found under the link in Ref. [6]. Once Public Safety networks based on LTE technology are deployed, new requirements may arise leading to new features standardized by 3GPP or other standardization organizations. It will be interesting to see how a standardized broadband wireless system such as LTE will help to build reliable Public Safety networks in a cost-efficient manner. These networks and their users will benefit from a radio/network technology and from devices such as Smartphones that have already changed the way people communicate with each other significantly over the last few years.

References

[1] European Union regulation: http://ec.europa.eu/enterprise/sectors/automotive/safety/ecall/index_en.htm.
[2] 3GPP TS 22.179: "3rd Generation Partnership Project; Technical Specification Group Services and System Aspects; Mission Critical Push to Talk MCPTT (Release 13)".
[3] OMA Push to Talk Over Cellular V2.1 Enabler: http://technical.openmobilealliance.org/Technical/technical -information/release-program/current-releases/poc-v2-1.
[4] 3GPP TR 22.897: "3rd Generation Partnership Project; Technical Specification Group Services and System Aspects; Study on Isolated Evolved Universal Terrestrial Radio Access Network (E-UTRAN) Operation for Public Safety (Release 13)".
[5] 3GPP TS 22.346: "3rd Generation Partnership Project; Technical Specification Group Services and System Aspects; Isolated E-UTRAN Operation for Public Safety (Release 13)".
[6] 3GPP Work Items on LTE for Public Safety (Authority-to-Authority) Communications: http://www.3gpp.org/ftp /information/work_plan/description_releases/Previous_versions/.

Appendix A

A.1 Call Flows

A.1.1 Attach

The User Equipment (UE) needs to register with the network to receive Evolved Packet System (EPS) services like Internet connectivity. This registration is referred to as network attachment, in short Attach. "Always-on" Internet Protocol (IP) connectivity is enabled in EPS by establishing a so-called default EPS bearer during Attach (see Figure A.1).

1. The UE initiates the Attach procedure by sending an Attach Request that is encapsulated in the Radio Resource Control (RRC) connection setup complete message. The RRC message is sent to the Evolved NodeB (eNodeB). The UE provides its Globally Unique Temporary Identity (GUTI) in the Attach, if this is available. If GUTI is not available, it provides its International Mobile Subscriber Identity (IMSI). If an UE without Universal Subscriber Identity Module (USIM) initiates the Attach procedure for emergency services, it may provide its International Mobile Equipment Identity (IMEI). The Attach Type can either be an Initial Attach, Handover, or Emergency. If the UE has valid security parameters, then the Attach Request is integrity protected.
2. The eNodeB forwards the Attach Request to the selected Mobility Management Entity (MME) (either derived from the old Globally Unique Mobile Management Entity Identifier (GUMMEI) or selected based on MME selection function) in a S1-MME control message.
3. If the UE context is not present anywhere in the network (in old or new MME) and the Attach Request was not integrity protected or if the integrity check failed, then authentication and Non Access Stratum (NAS) security setup to activate integrity protection and NAS ciphering are mandatory. Otherwise it is optional during Attach procedure. MME initiates this procedure for authenticating the UE and involves Home Subscriber Server (HSS) to obtain authentication vectors for the UE.
4. The MME sends an Update Location Request message to the HSS to update the records for the given IMSI. This message is sent, if any of the following conditions is true:
   a. MME has changed since the last detach.
   b. There is no valid subscription context for the UE in the MME.
   c. ME identity has changed, and UE provides an IMSI.
   d. UE provides an old GUTI which does not refer to a valid context in the MME or.
   e. UE is not performing an emergency attach.
5. The HSS acknowledges the Update Location message with an Update Location Ack.

LTE for Public Safety, First Edition. Rainer Liebhart, Devaki Chandramouli, Curt Wong and Jürgen Merkel.
© 2015 John Wiley & Sons, Ltd. Published 2015 by John Wiley & Sons, Ltd.

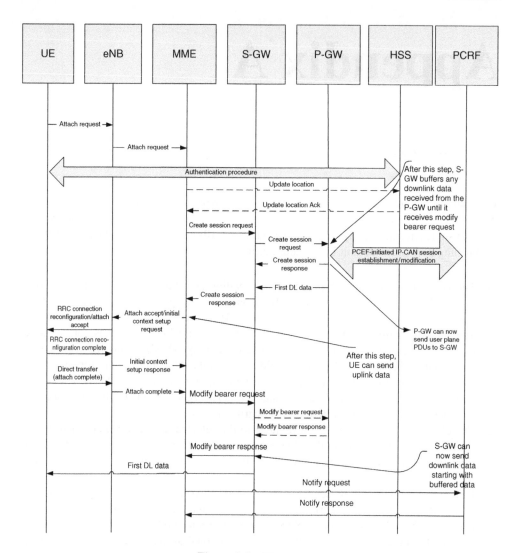

Figure A.1   EPS Attach

6. The MME selects an Serving Gateway (S-GW) and Packet Data Network Gateway (P-GW) and allocates an EPS Bearer Identity for the default bearer associated with the UE. Then it sends a Create Session Request to the selected S-GW to initiate the creation of default bearers for the UE.

7. The S-GW creates a new entry in its EPS Bearer table and sends a Create Session Request to the P-GW.

8. If dynamic policy control (PCC, Policy and Charging Control) is deployed and the handover indication is not present, the P-GW performs IP-Connectivity Access Network (IP-CAN) Session Establishment procedure and thereby obtains the default PCC rules for the UE. If PCC is deployed and the handover indication is present, the P-GW reports

the new access type to the Policy and Charging Rules Function (PCRF). If PCC is not deployed, the P-GW may apply local Quality of Service (QoS) policies. Either of these conditions could result in establishment of a number of dedicated bearers for the UE in association or combination with the default bearer.

9. The P-GW creates a new entry in its EPS bearer context table and generates a General Packet Radio Service (GPRS) Charging Correlation Identifier. This allows the P-GW to route user plane packets between the S-GW and the Packet Data Network (PDN) (e.g., the Internet), and to start a new charging session. The P-GW responds to the S-GW with Create Session Response. If the Handover Indication is present, the P-GW does not yet send downlink packets to the S-GW until the downlink path is switched. The S-GW responds to the MME with Create Session Response.

10. The MME determines the UE AMBR to be used by the eNodeB based on the subscribed UE-AMBR and the APN Aggregate Maximum Bit Rate (APN-AMBR) for the default Access Point Name (APN). The MME responds to the UE with Attach Accept. The GUTI is included, if the new MME allocates a new GUTI. This message is contained in the S1-MME control message Initial Context Setup Request. This S1 control message also includes the AS security context information for the UE, the Handover Restriction List, the EPS Bearer QoS, the UE-AMBR, EPS Bearer Identity, as well as the Tunnel Endpoint Identifier (TEID) and the address of the S-GW for user plane traffic.

11. The eNodeB sends the RRC Connection Reconfiguration message including the EPS Radio Bearer Identity along with the Attach Accept to the UE. The APN for which the default bearer is associated is provided to the UE. IP address allocation procedure for the UE is initiated during the attach procedure.

12. The UE sends the RRC Connection Reconfiguration complete message to the eNodeB and the eNodeB sends the Initial Context Response message to the MME. This Initial Context Response message includes the TEID of the eNodeB and the address of the eNodeB used for downlink traffic on the S1-U reference point.

13. The UE sends a Direct Transfer message to the eNodeB, which includes the Attach Complete.

14. The eNodeB forwards the Attach Complete message to the new MME in an Uplink NAS Transport message. The UE can now send uplink data packets toward the eNodeB, which will then be tunneled to the S-GW and P-GW.

15. The MME sends a Modify Bearer Request message to the S-GW. This includes the EPS bearer ID, eNodeB address, and the eNodeB TEID.

16. The S-GW sends Modify Bearer Request toward P-GW only when UE is connected via non-3GPP (3rd Generation Partnership Program) access. This can be determined based on the presence of the Handover Indication. If the Handover Indication is present, the S-GW sends a Modify Bearer Request message to the P-GW to prompt the P-GW to tunnel packets from non-3GPP IP access to 3GPP access system and immediately start routing packets to the S-GW for the default and any dedicated EPS bearers established.

17. The P-GW sends a Modify Bearer Response to the S-GW and a Modify Bearer Response to the MME. Bearers are now established and ready to exchange uplink and downlink packets.

18. The MME sends a Notify Request to the HSS to update the HSS with APN and P-GW pair. The HSS stores the APN and the P-GW identity pair and sends a Notify Response to the MME.

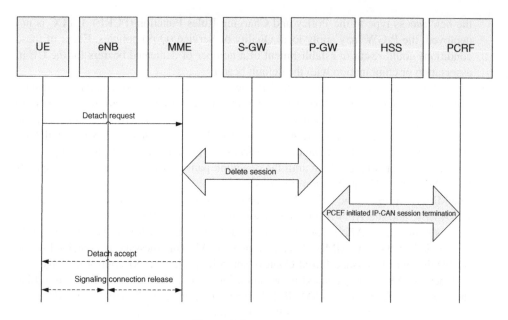

Figure A.2   EPS Detach

For further details, please refer to 3GPP TS 23.401 [1].

A.1.2   Detach

The Detach procedure allows the user equipment to inform the network that it does not want to access the network any longer and allows the network to inform the user equipment that it does not have access to the network any longer (see Figure A.2).

1. The UE initiates the Detach procedure by sending a Detach Request to the MME and also indicates, if the detach is due to power off.
2. The MME triggers deletion of established sessions by sending Delete Session Request to the S-GW, which in turn sends it to the P-GW to deactivate all the bearers for the UE. Both P-GW and S-GW release the resources and acknowledge the Delete Session Request.
3. The P-GW employs a Policy Charging Enforcement Function (PCEF) initiated IP-CAN Session Termination Procedure with the PCRF to indicate that EPS bearers are released, if PCC is deployed in the network.
4. If the Detach is not due to a switch off, the MME sends Detach Accept to the UE.
5. The MME releases the S1-MME signaling connection for the UE by sending S1 Release Command to the eNodeB with cause value set to Detach. This triggers the release of logical S1-AP signaling connections over the S1-MME interface and all S1 bearers on S1-U for a UE. The UE moves from ECM-CONNECTED to ECM-IDLE state in both UE and MME. UE-related context information is deleted in the eNodeB.

For further details, please refer to 3GPP TS 23.401 [1].

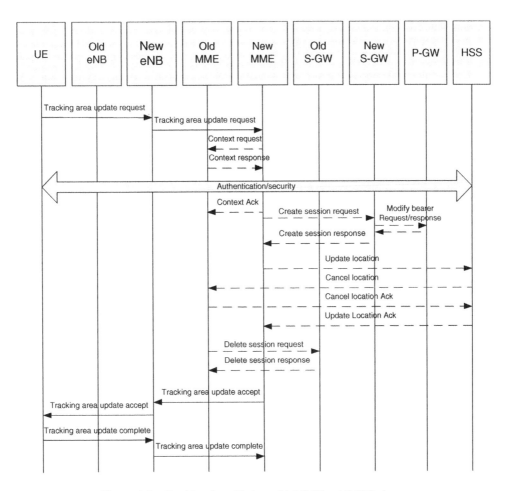

Figure A.3   Tracking Area Update with MME and S-GW change

A.1.3   Tracking Area Update

Once the UE is successfully attached, it needs to keep the network informed about its current location in order to be reachable for downlink signaling and user data even when it is in idle mode, that is, when the signaling connection to the UE has been released. In order to keep the network informed, the UE performs a Tracking Area Update (TAU) (see Figure A.3).

1. The UE sends a TAU request encapsulated within an RRC message to the new eNB, for example, because the tracking area has changed or its periodic TAU timer has expired. The TAU contains information such as old GUTI, last visited Tracking Area Identity (TAI), and EPS bearer status.
2. The eNodeB derives the MME from the RRC parameters such as GUMMEI and the selected network. eNodeB forwards the TAU request to the new MME.
3. The new MME uses the GUTI to derive the old MME and sends a Context Request message to the old MME to retrieve user context information. The new MME provides the

complete NAS message TAU Request, including the integrity protection information, with the GPRS Tunneling Protocol (GTP) message Context Request to the old MME to check the integrity of the message. The old MME answers with a Context Response including UE's context data, including the EPS security context, and mobility management and session management information. If the old MME indicates in the Context Response message that the integrity check was not successful, the new MME will authenticate the UE.

4. In this case the new MME decides to relocate the S-GW, for example, because the old S-GW cannot serve the UE any longer. The new MME sends a Context Acknowledge with S-GW change indication to the old MME.

5. The new MME sends a Create Session Request with IMSI, bearer context data, P-GW address, MME Address for each PDN connection to the new S-GW.

6. The new S-GW informs the P-GWs about, for example, a radio access type change (could be Wireless Local Area Network (WLAN) to Evolved Universal Terrestrial Radio Access Network (E-UTRAN)) by sending Modify Bearer Request per PDN connection to the P-GWs. The P-GWs update their bearer contexts and send a Modify Bearer Response.

7. The new S-GW returns a Create Session Response with its own address, user plane, and control plane connection addresses to the new MME.

8. If the new MME has no subscription data for the UE, it sends an Update Location Request to the HSS. MME registration is updated in the HSS, that is, MME address is stored in the HSS as serving node for the UE. In turn the HSS sends Cancel Location to the old MME and an Update Location Ack including the subscription data to the new MME.

9. The old MME deletes bearer resources by sending Delete Session Request to the old S-GW.

10. Finally, the new MME sends a TAU accept with GUTI, TAI list, and EPS bearer status to the UE. If the MME was changed, the new MME will also include a new GUTI which identifies the new MME. If a new GUTI was allocated, the UE has to confirm the receipt of the identity with a NAS message TAU Complete.

For further details, please refer to 3GPP TS 23.401 [1].

A.1.4   Paging

The paging procedure (see Figure A.4) is initiated by the network to request the establishment of a NAS signaling connection to the UE. When the UE is in RRC-IDLE, the UE monitors for paging messages and the frequency for this monitoring is determined by the "DRX cycle in idle mode".

1. The P-GW receives a downlink IP packet for the UE from an external source (e.g., an Application Server).

2. The P-GW forwards the downlink IP packet to the corresponding S-GW. The route for the packet is determined by flow attributes such as packet filters, Downlink Traffic Flow Template (DL TFT), and TEID.

3. The S-GW receives the packet and determines that the corresponding S1-U tunnel is not established. Thus, it buffers the packet and determines from its context data the corresponding MME that is serving the UE. It sends a Downlink Data Notification (DDN) to the MME that has control plane connectivity for the given UE. The Address Resolution

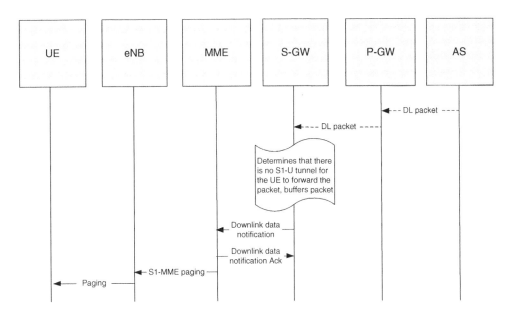

Figure A.4   Paging triggered by a DL packet

Protocol (ARP) and EPS Bearer ID are always set in Downlink Data Notification. The MME responds with a Downlink Data Notification Ack message. The priority indicator, that is ARP, is derived from the bearer triggering the Downlink Data Notification.

4. The MME receives the Downlink Data Notification message. It determines whether the UE is registered in the MME. The MME uses parameters such as EPS bearer ID, ARP to determine the paging strategy (i.e., paging in all TAIs of the stored TAI list, stepwise paging by initially paging only in the last known eNB or last known TAI, etc.). Accordingly, it initiates a paging message toward the eNB over S1-AP.
5. The eNB initiates paging based on UE's idle mode paging DRX cycle.

For further details, please refer to 3GPP TS 23.401 [1] and 3GPP TS 24.301 [2].

A.1.5   Service Request

The Service Request can be sent by the UE either based on pending UL data or signaling messages or in response to a paging message from the network. Service Request (see Figure A.5) is normally sent to request the establishment of a NAS signaling connection.

1. The UE sends a NAS message Service Request toward the MME. The NAS message is encapsulated in an RRC message sent to the eNB.
2. The eNB forwards the NAS message to the MME. The NAS message is encapsulated in an S1-AP Initial UE message. The Initial UE message also includes the TAI and ECGI of the cell and the S-TMSI. The MME can use the S-TMSI to locate the UE context.
3. NAS authentication/security procedures may be performed. This is an optional step during a Service Request procedure.

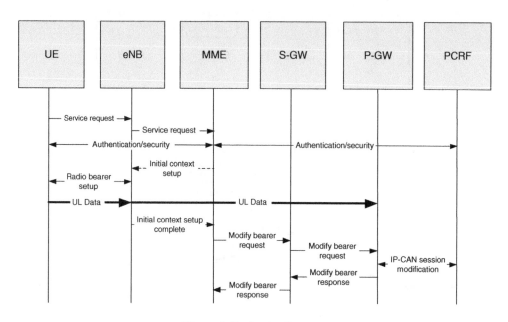

Figure A.5   Service Request

4. The MME sends a S1-AP Initial Context Setup Request to the eNB. It contains, for example, S-GW address, S1-TEID(s), EPS Bearer QoS(s), Security Context, and MME Signaling Connection ID. This step activates the radio and S1 bearers for all active EPS bearers. The eNB stores the Security Context, MME Signaling Connection Id, EPS Bearer QoS(s), and S1-TEID(s) in the UE context.

5. The eNB performs the radio bearer establishment procedure. The user plane security is established at this step. When the user plane radio bearers are setup, EPS bearer state synchronization is performed between UE and network, that is, the UE shall locally remove any EPS bearers for which no radio bearers are setup and, if the radio bearer for a default EPS bearer is not established, the UE shall locally deactivate all EPS bearers associated to that default EPS bearer.

6. The UL data from UE can now be forwarded by the eNB to the S-GW and P-GW.

7. The eNodeB sends an S1-AP Initial Context Setup Complete to the MME. It includes the eNB address, S1 TEID(s) for the DL, list of accepted EPS bearers, and list of rejected EPS bearers.

8. The MME sends a Modify Bearer Request per PDN connection to the S-GW. This includes all the parameters such as S1 TEID(s) provided by the eNB. The S-GW is now able to transmit DL data toward the UE. If a default EPS bearer is not accepted by the eNB, all the EPS bearers associated to that default bearer shall be treated as non-accepted bearers. The MME releases the non-accepted bearers by triggering the bearer release procedure.

9. If a RAT type change was requested by the P-GW and the RAT type has changed, the S-GW shall send the Modify Bearer Request per PDN connection to the P-GW. If the S-GW receives a DL packet for a non-accepted bearer, the S-GW drops the DL packet and does not send a Downlink Data Notification to the MME.

10. If dynamic PCC is deployed, the P-GW interacts with the PCRF to get the PCC rule(s) according to the Radio Access Type (RAT) type. If dynamic PCC is not deployed, the P-GW may apply local QoS policies.
11. The P-GW sends the Modify Bearer Response to the S-GW.
12. The S-GW sends the Modify Bearer Response to the MME.

For further details, please refer to 3GPP TS 23.401 [1] and 3GPP TS 24.301 [2].

A.1.6   X2-Based Handover

If both eNodeBs are connected to the same MME, source eNodeB and target eNodeB can exchange the S1AP signaling for the handover preparation and execution directly via the X2 interface. In this case we are talking about X2-based handover (see Figure A.6).

1. Once the source eNodeB determines that a handover is required (e.g., based on signal strength measurements), it initiates the so-called handover preparation procedure. Within this procedure, the source eNodeB sends a handover command to the UE. During handover execution, user data can be forwarded from the source eNodeB to the target eNodeB without involving the EPC.
2. After handover execution, the target eNodeB sends a Path Switch request to MME to inform that the UE has changed cell, including TAI and ECGI of the target cell and the list of EPS bearers.
3. The MME determines that the S-GW can serve the UE and sends a Modify Bearer Request to the S-GW for each PDN connection that has been accepted by the target eNodeB.
4. The MME releases dedicated bearers that were not accepted. If the S-GW receives a DL packet for a non-accepted bearer, it drops the packet. If the default bearer of a PDN connection has not been accepted by the target eNodeB and multiple PDN connections are active, the MME releases the whole PDN connection for which the default bearer was not accepted.
5. If required, the S-GW sends a Modify Bearer Request to the P-GWs to inform them about changed user location or time zone. This message is sent per PDN connection.
6. The S-GW sends DL packets to the target eNodeB. A Modify Bearer Response is sent to the MME.
7. To allow reordering of packets in the target eNodeB, the S-GW sends "end marker" packets on the old path immediately after switching the path.
8. The MME confirms the Path Switch Request with the Path Switch Request Ack. If the UE-AMBR is changed, the MME provides the updated value of UE-AMBR to the target eNodeB in the Path Switch Request Ack message. If some bearers have not been switched, the MME shall indicate the failed bearers in the Path Switch Request Ack. Resources for failed dedicated bearers are released. The target eNodeB also removes corresponding bearer context data.
9. The target eNodeB sends Release Resource to the source eNodeB.
10. Finally, the UE might send a TAU request to the new eNodeB depending on the TAI and other TAU triggering criteria.

For further details, please refer to 3GPP TS 23.401 [1] and 3GPP TS 36.300 [3].

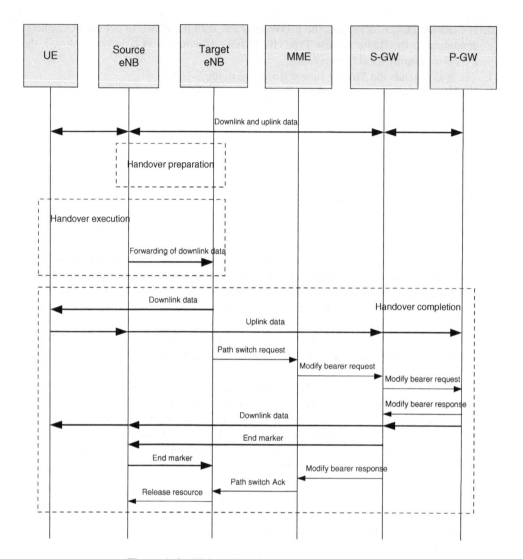

Figure A.6   X2-based handover without S-GW change

A.1.7   S1-Based Handover

S1-based handover is used in cases when X2-based handover cannot be used, for example, when source eNodeB and target eNodeB are served by different MMEs or have no direct X2 connection. During the handover preparation (see Figure A.7), the signaling information is exchanged between source eNodeB and target eNodeB via the involved MME(s), that is, messages are sent via the S1 interfaces, and possibly via the S10 interface between the MME(s).

1. The source eNodeB decides to initiate a S1-based handover to the target eNodeB, for example, as there is no X2 connection to the target eNodeB.

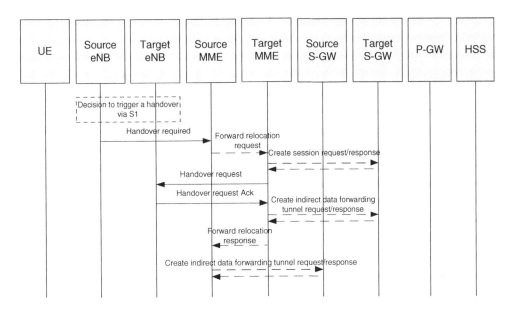

Figure A.7   S1-based handover – preparation phase

2. The source eNodeB sends Handover Required including target eNodeB identity and target TAI to the source MME. The target TAI is used by the source MME to select a target MME.

3. If the MME is relocated, the source MME sends Forward Relocation Request with the MME UE context and target TAI to the target MME. The target TAI is used by the target MME to determine whether S-GW relocation is needed and to select the new S-GW. The MME UE context includes IMSI, ME identity, UE security context, AMBR, S-GW address, and EPS bearer context data (e.g., P-GW addresses, APN, and S-GW addresses). If the MME has been relocated, the target MME decides whether the S-GW has to be relocated, otherwise the source MME decides whether S-GW needs to be relocated.

4. If a new S-GW is selected, the target MME sends a Create Session Request per PDN connection to the target S-GW. The target S-GW allocates the S-GW addresses and TEIDs for the uplink traffic and sends a Create Session Response to the target MME.

5. The Target MME sends a Handover Request with EPS Bearers to setup and AMBR to the target eNodeB. This message creates the UE context in the target eNodeB, including information about the bearers and the security context. For each EPS bearer to be setup, it includes S-GW address and uplink TEID for the user plane and the bearer QoS.

6. The target eNodeB sends a Handover Request Acknowledge with a list of EPS bearers to be setup and EPS bearers failed to setup to the target MME. The EPS Bearer setup list includes a list of addresses and TEIDs allocated at the target eNodeB for downlink traffic on the S1-U reference point and addresses and TEIDs for receiving forwarded data if necessary. If the UE-AMBR is changed, the MME shall recalculate the new UE-AMBR and signal the modified UE-AMBR value to the target eNodeB. If none of the default EPS bearers have been accepted by the target eNodeB, the target MME rejects the handover.

7. If indirect forwarding applies and the S-GW is relocated, the target MME sets up forwarding parameters by sending Create Indirect Data Forwarding Tunnel Request to the S-GW. The

S-GW sends a Create Indirect Data Forwarding Tunnel Response to the target MME. If the S-GW is not relocated, indirect forwarding can be set up later on.

8. If the MME has been relocated, the target MME sends a Forward Relocation Response to the source MME. For indirect forwarding, this message includes the S-GW address and TEIDs for indirect forwarding.

9. If indirect forwarding applies, the source MME sends Create Indirect Data Forwarding Tunnel Request to the S-GW. If the S-GW is relocated, it includes the tunnel identifier to the target S-GW. The S-GW responds with Create Indirect Data Forwarding Tunnel Response to the source MME.

Procedure for the S1 HO execution phase (Figure A.8):

1. Once the handover preparation phase is completed, the source MME sends a Handover Command including list of bearers that are subject to forwarding and bearers to release to the source eNodeB. The list of bearers that are subject to forwarding includes a list of addresses and TEIDs allocated for forwarding. The Handover Command is sent to the UE. Upon reception of this message, the UE will remove any EPS bearers for which it did not receive the corresponding EPS radio bearers in the target cell.

2. The source eNodeB may send Status Transfer to the target eNodeB via the MME to convey the status of the radio bearers. In case of MME relocation, the source MME sends this information to the target MME via Forward Access Context Notification. The source or target MME sends the information then to the target eNodeB.

3. The source eNodeB forwards downlink data received by the source eNodeB towards the target eNodeB for bearers subject to data forwarding.

4. After the UE has synchronized to the target cell, it sends a Handover Confirm to the target eNodeB. Downlink packets forwarded from the source eNodeB are sent to the UE. Uplink packets are forwarded to the target S-GW and to the P-GW. The target eNodeB sends a Handover Notify with TAI and ECGI to the target MME.

5. If the MME has been relocated, the target MME sends a Forward Relocation Complete Notification to the source MME. The source MME sends a Forward Relocation Complete Acknowledge to the target MME.

6. The MME sends a Modify Bearer Request including eNodeB address and TEID allocated for downlink traffic for the accepted bearers to the target S-GW for each PDN connection, including the PDN connections that need to be released. If the S-GW supports Modify Access Bearers Request procedure and if there is no need for the S-GW to send signaling messages to the P-GW (e.g., user location or time zone data required for charging), the MME may send Modify Access Bearers Request per UE to the S-GW.

7. The MME releases the non-accepted dedicated bearers. If the S-GW receives a DL packet for a non-accepted bearer, it drops the packet.

8. If the default bearer of a PDN connection has not been accepted by the target eNodeB and there are other PDN connections active, the MME releases this connection.

9. If the S-GW is relocated, the target S-GW assigns addresses and TEIDs per bearer for downlink traffic. It sends a Modify Bearer Request including its addresses for user plane and TEIDs per PDN connection to the P-GWs. The S-GW allocates downlink TEIDs on S5/S8 interfaces also for non-accepted bearers. The P-GW updates its context data and returns a Modify Bearer Response with charging ID and MSISDN to the target S-GW.

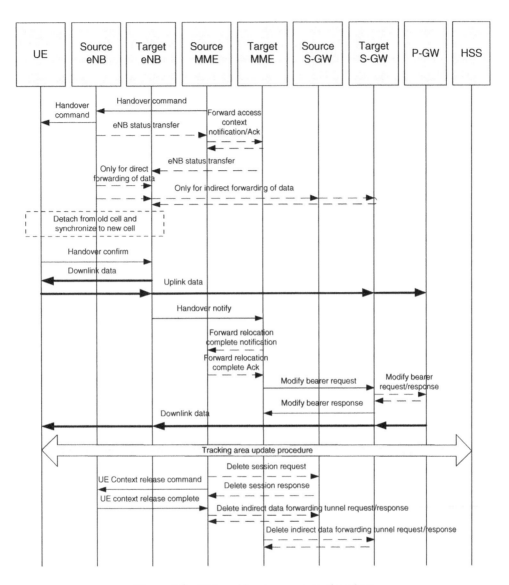

Figure A.8   S1-based handover – execution phase

The P-GW sends downlink packets to the target S-GW, which will be forwarded to the target eNodeB. If the S-GW is not relocated, it sends also necessary information such as user location or time zone to the P-GWs by a Modify Bearer Request. This message is answered by the P-GW with a Modify Bearer Response. If the S-GW is relocated, the P-GW sends one or more "end marker packets" on the old path to allow the reordering of packets in the target eNodeB. The source S-GW forwards "end marker packets" to the source eNodeB.

10. The S-GW returns a Modify Bearer Response or a Modify Access Bearers Response with its address and TEIDs for uplink traffic to the MME. If the S-GW does not change, it sends one or more "end marker packets" on the old path after switching the path.
11. The UE may initiate a TAU. The target MME performs only a subset of the TAU procedure, for example, context data are not transferred between source and target MME.
12. The source MME sends a UE Context Release Command to the source eNodeB. The source eNodeB releases its resources related to the UE and responds with a UE Context Release Complete. If the source MME received the S-GW change indication in the Forward Relocation Response message, it deletes the EPS bearer resources by sending Delete Session Request to the source S-GW. The source S-GW acknowledges with Delete Session Response.
13. If indirect forwarding was used, the source MME sends a Delete Indirect Data Forwarding Tunnel Request to the (source and/or target) S-GW to release temporary resources.

For further details, please refer to 3GPP TS 23.401 [1] and 3GPP TS 36.300 [3].

A.1.8   MBMS Session Start

The BM-SC initiates the MBMS Session Start procedure (see Figure A.9) when it is ready to send DL data. This is a request to activate all necessary bearer resources in the network for the transfer of MBMS data and to notify interested UEs of the imminent start of the data transmission.

1. BM-SC sends a Session Start Request toward MBMS GW(s) to initiate MBMS sessions. It also indicates the impending transmission of downlink data and provides the attributes that are relevant for the session. Some key attributes are TMGI, Flow ID, QoS, MBMS service area, duration of the session, time to MBMS data transfer, and MBMS data transfer start. In addition, BM-SC also provides a list of MME(s) that are relevant for the session. The MBMS GW responds with a Session Start Response message.
2. The MBMS GW creates an MBMS bearer context for the session. It sends a Session Start Request including the session attributes received from the BM-SC. In addition, it also includes transport network IP multicast address, IP address of the multicast source, and GTP TEID to the list of MMEs that were provided by the BM-SC. Data are transmitted to eNB(s) using IP multicast. This avoids replication of user data in multiple GTP tunnels toward all the eNB(s) in the service area.
3. MME creates an MBMS context for the session. It sends a Session Request with all the session attributes received from the MBMS GW. It can either send the Session Request to all the connected Multicell Coordination Entity (MCE)(s) or only to certain MCE(s) based on service area.
4. The E-UTRAN network establishes the necessary radio resources for the transfer of MBMS data to the interested UEs. MCE checks whether the radio resources are sufficient for the establishment of new MBMS service(s) in the area it controls. If resources are available, MCE sends MBMS Session Start to the eNBs in the targeted service area. If not, MCE decides not to establish the radio bearers of the MBMS service(s) and does not forward the MBMS Session Start Request to the involved eNBs. It may also decide to preempt radio resources from other radio bearer(s) of ongoing MBMS service(s) according to the ARP. The MCE confirms the reception of the MBMS Session Start Request to the MME.

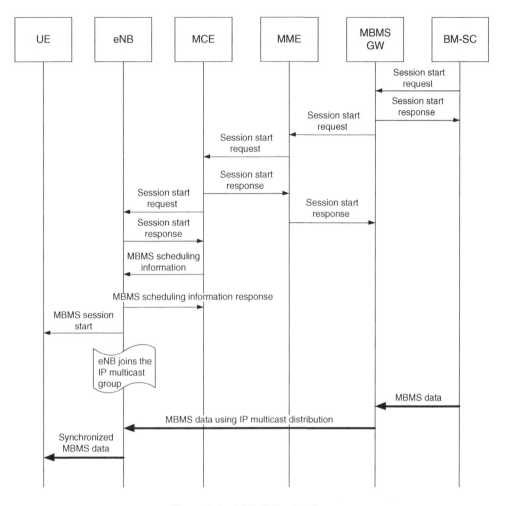

Figure A.9   MBMS Session Start

5. MCE sends the MBMS Session Start to the eNBs in the targeted MBMS service area. eNBs confirm the reception of the MBMS Session Start message.
6. MCE sends the MBMS scheduling information message to the eNBs including the updated Multicast Control Channel (MCCH) information, which carries MBMS service's configuration information. eNBs confirm the reception of this message.
7. The eNB indicates MBMS Session Start to UEs by MCCH change notification and updated MCCH information, which carries MBMS service's configuration information. eNB joins the IP multicast group to receive the downlink user plane data from MBMS GW.
8. The BM-SC starts sending the MBMS data toward MBMS-GW. MBMS GW sends the data using IP multicast distribution toward all joined eNBs.
9. The eNB transmits MBMS user data in a synchronized manner to all the interested UE(s).

For further details, please refer to 3GPP TS 23.246 [4] and 3GPP TS 36.300 [3].

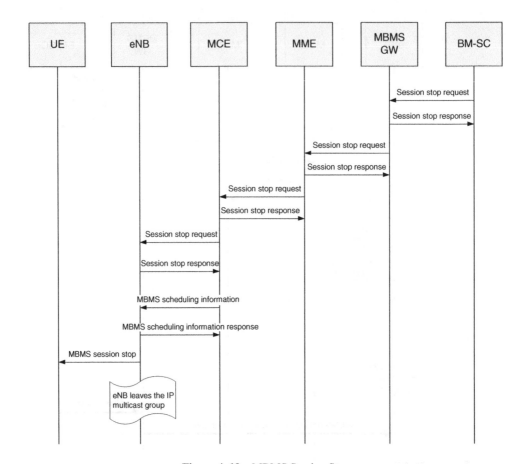

Figure A.10   MBMS Session Stop

A.1.9   MBMS Session Stop

The BM-SC initiates the MBMS Session Stop procedure (see Figure A.10) when it considers the MBMS session to be terminated.

1. The BM-SC sends a Session Stop Request to the MBMS GW(s) to indicate the end of a session and that the bearer plane resources can be released. The MBMS context can be uniquely identified by the TMGI and Flow ID. BM-SC sets the status of the MBMS Bearer Context to "Standby."
2. The MBMS GW releases the MBMS bearer context in case of a broadcast MBMS bearer service and responds to the BM-SC with Session Stop Response.
3. The MBMS GW forwards Session Stop Request to the MME(s) to which it previously sent the Session Start Request and it sets the status of its MBMS bearer context to "Standby."
4. The MME sets the status of its MBMS bearer context to "Standby" and responds to the MBMS GW with Session Stop Response.
5. The MME forwards Session Stop Request to the MCE(s) to which it previously sent the Session Start Request.

6. Each MCE responds with Session Stop Response to the MME. The MME releases the MBMS bearer context.
7. Each MCE forwards the Session Stop to the eNBs. The eNB confirms the reception of the MBMS Session Stop.
8. The MCE sends the MBMS scheduling information message to the eNBs including the updated MCCH information, which carries MBMS service's configuration information. The eNB confirms the reception of this message.
9. The eNB indicates MBMS session stop to UEs by removing any service configuration associated with the stopped session from the updated MCCH message. The corresponding E-RAB is released, and eNB leaves the IP multicast group.

For further details, please refer to 3GPP TS 23.246 [4] and 3GPP TS 36.300 [3].

A.1.10   MBMS Session Update

The BM-SC initiates a MBMS Session Update procedure (see Figure A.11) when attributes such as service area or bearer ARP value for an ongoing MBMS session have to be modified. This procedure can be used to notify eNBs to join or leave a service area or change the priority of an ongoing group communication service.

1. The BM-SC sends a Session Update Request to the MBMS GW, which includes information such as TMGI, Flow Identifier, session identifier, QoS, service area, estimated session duration, time to data transfer, data transfer start, and a list of MBMS control plane nodes. TMGI and session identifier are used to identify the ongoing session. Except the ARP parameter, all other QoS parameters should be identical as in the MBMS Session Start Request. The ARP parameter may be different if it needs to be updated. Service area and list of MBMS control plane nodes can define the new MBMS service area. The estimated session duration shall be set to a value corresponding to the remaining part of the session. The MBMS GW sends a Session Update Response to the BM-SC.
2. The MBMS GW sends an MBMS Session Start Request to newly added MMEs, an MBMS Session Stop Request to removed MMEs, and an MBMS Session Update Request to all other MMEs in the control plane node list.
3. The MME sends an MBMS Session Update Request including the new session attributes received from the MBMS GW. It can either send the Session Update Request to all the connected MCE(s) or only to certain MCE(s) based on service area. MCE(s) eventually forward it to the eNB(s) that belong to the service area. If at least one node newly added to the MBMS service area accepts the Session Update Request and the proposed IP multicast and source address for backbone distribution, the MME includes an indication that IP multicast distribution is accepted in the MBMS Session Update Response to MBMS GW. If at least one of the nodes does not accept these addresses, the MME uses normal point-to-point MBMS bearer establishment procedure for these nodes and responds with an MBMS Session Update Response providing the tunnel identifier for bearer plane that the MBMS GW has to use.
4. If the E-UTRAN network has no MBMS bearer context with the TMGI indicated in the MBMS Session Update Request, it creates an MBMS bearer context. Otherwise the E-UTRAN updates the existing context. E-UTRAN responds to the MME with a Session

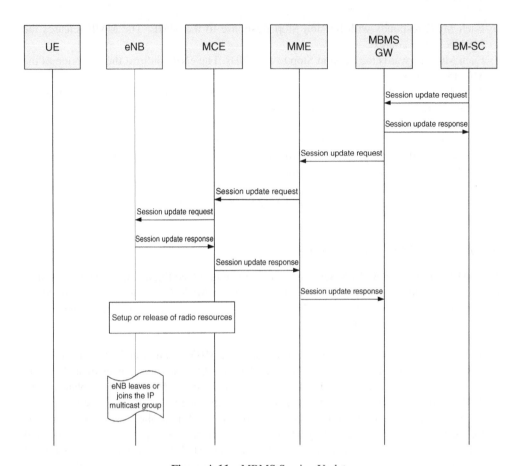

Figure A.11   MBMS Session Update

Update Response. The MME updates the session attributes in its MBMS bearer context and responds to the MBMS GW.

5. The E-UTRAN network establishes or releases radio resources for the transfer of MBMS data. If the ARP parameter is updated, the MCE shall ensure that any necessary changes to radio resources are synchronized across all eNodeBs in the corresponding MBSFN area. For E-UTRAN, the radio resource set up is scheduled using the MBMS data transfer start parameter, if present, otherwise using the time to MBMS data transfer parameter, if present.

6. The eNodeBs send IP multicast Join or Leave messages to the received user plane IP multicast address allocated by the MBMS GW.

For further details, please refer to 3GPP TS 23.246 [4] and 3GPP TS 36.300 [3].

A.1.11   UE-requested PDN Connectivity

EPS supports simultaneous exchange of IP traffic via multiple PDNs through the use of separate P-GWs or a single P-GW. This is controlled by operator policies and defined in the user

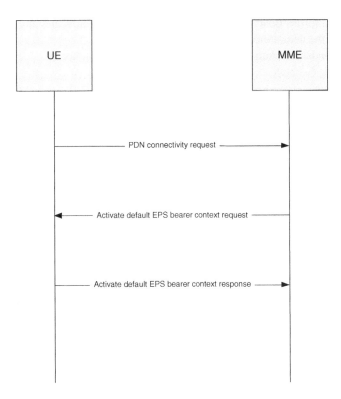

Figure A.12   UE requested PDN Connectivity

subscription. Hence, EPS supports UE-initiated connectivity establishment in order to allow multiple PDN connections to one or more PDNs. This procedure may trigger establishment of dedicated EPS bearers(s) for that UE. During the attach procedure, PDN connectivity request is piggybacked within the attach request. If the UE is requesting for an additional PDN connection, this can be performed after successful normal Attach and UE can send a stand-alone PDN connectivity request message. In this case, the Activate EPS Default Bearer Context procedure is initiated in response to the UE requested PDN Connectivity request message (see Figure A.12).

1. The UE initiates additional PDN Connectivity request procedure by sending PDN Connectivity Request. It specifies the type of PDN requested. If it is requesting PDN connectivity for emergency services, this is specified in the request type.
2. If the MME receives a PDN Connectivity Request from an emergency attached UE or the PDN Connectivity Request is for normal services and the mobility or access restrictions do not allow the UE to access normal services, the MME rejects this request. If the PDN connection is requested for emergency services by a normally attached UE, the MME selects a P-GW from the VPLMN based on APN or the statically configured P-GW from the emergency configured data stored within the MME. Upon selection of S-GW and P-GW, MME initiates the session creation for the UE with S-GW and P-GW and responds to the UE with

Activate Default EPS Bearer Context Request. The network allocates an IP address for the UE as part of the default bearer activation procedure.
3. The UE responds with Activate Default EPS Bearer Context Accept.

For further details, please refer to 3GPP TS 23.401 [1] and 3GPP TS 24.301 [2].

A.1.12   Dedicated Bearer Context Activation

With this procedure the network can establish a dedicated bearer for a UE fulfilling special QoS requirements. As an example, the network can detect establishment of a VoLTE call and decide to establish a dedicated voice bearer (see Figure A.13).

1. If dynamic PCC is deployed, IP-CAN session modification procedure can be a trigger for the dedicated bearer activation procedure.
2. The P-GW uses the QoS policy to assign the EPS Bearer QoS. It sends a Create Bearer Request to the S-GW.
3. The S-GW sends Create Bearer Request to the MME. If the UE is in ECM-IDLE state, the MME will trigger the Network Triggered Service Request procedure in which case steps 4–7 are either combined in to that procedure or performed stand alone.
4. The MME selects an EPS Bearer Identity, which has not yet been assigned to the UE. The MME builds a Session Management Request. The MME includes the EPS bearer identity of the associated default bearer as the linked EPS bearer identity. The MME signals the Bearer Setup Request which includes the Session Management request message to the eNodeB.
5. The eNodeB maps EPS Bearer QoS to Radio Bearer QoS. It then sends a RRC Connection Reconfiguration message to the UE. The UE stores the EPS bearer identity. The linked EPS bearer identity included in the Session Management request indicates to the UE to which default bearer, IP address, and PDN the dedicated bearer is linked to.
6. The UE acknowledges the radio bearer activation to the eNodeB with a RRC connection reconfiguration complete message.
7. The eNodeB acknowledges the bearer activation to the MME with Bearer Setup Response. The eNodeB indicates whether the requested EPS Bearer QoS can be allocated or not.
8. The UE builds a Session Management Response including EPS Bearer Identity. The UE then sends a Direct Transfer including Session Management Response to the eNodeB.
9. The eNodeB sends an Uplink NAS Transport message to the MME.
10. Upon receipt of Bearer Setup Response (step 7) and Session Management Response (step 9), the MME acknowledges the bearer activation to the S-GW by sending a Create Bearer Response.
11. The S-GW acknowledges the bearer activation to the P-GW by sending a Create Bearer Response.
12. If the dedicated bearer activation procedure was triggered by the PCRF, the P-GW indicates to the PCRF whether the requested QoS policy could be enforced or not, completing the PCRF-Initiated IP-CAN Session Modification procedure.

For further details, please refer to 3GPP TS 23.401 [1] and 3GPP TS 24.301 [2].

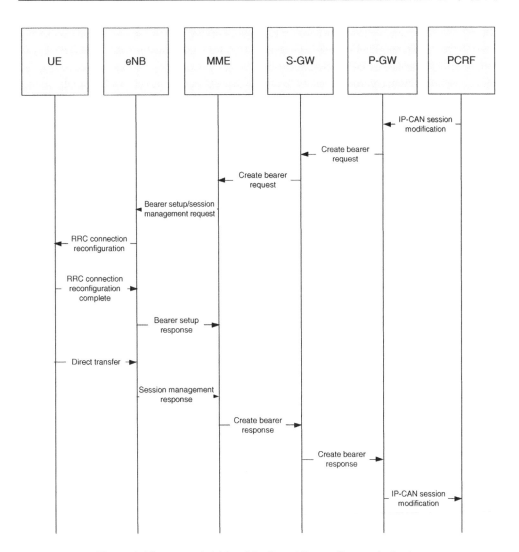

Figure A.13   Network-initiated Dedicated Bearer Context Activation

A.2   3GPP Reference Points

The list contains reference points used in this book.

Bp   Reference point for CDR file transfer from the CGF to the Billing Domain. For
                details, see 3GPP TS 32.251 [28].

Cx   Reference point between Call Session Control Function (CSCF) and HSS based
                on DIAMETER. This reference point is, for example, used to authenticate and
                authorize an IMS subscriber. For details, see 3GPP TS 29.228 [16] and 3GPP
                TS 29.229 [17].

Ga   Reference point between, for example, Serving GPRS Support Node (SGSN) and Charging Gateway Function (CGF) for CDR transfer based on GTP′. For details, see 3GPP TS 32.295 [29].

GC1   Reference point between UE and GCS Application server to allow application level control signaling such as group management and floor control, also for relaying any MBMS-specific bearer configuration data received from the BM-SC. This is not yet specified in 3GPP. For details, see TS 23.468 [35].

Gm   Reference between UE and P-CSCF to exchange Session Initiation Protocol (SIP) signaling messages. For details, see 3GPP TS 23.228 [5].

Gx   It provides transfer of QoS, policy, and charging rules from PCRF to PCEF located in the P-GW and is based on DIAMETER. QoS rules are provided only in case GTP is used at S5 (see also Gxc). For details, see 3GPP TS 29.212 [12].

Gxa   It provides transfer of QoS rules from PCRF to the trusted non-3GPP accesses system. It is based on DIAMETER. For details, see 3GPP TS 29.212 [12].

Gxb   It provides transfer of PCC rules from PCRF to ePDG. This reference point is not specified so far.

Gxc   It provides transfer of QoS rules from PCRF to S-GW in case PMIP is used for S5 interface as in such scenarios the bearers are terminated in the S-GW instead of P-GW. It is based on DIAMETER. For details, see 3GPP TS 29.212 [12]. Gxc is obsolete when GTP is used for S5 interface.

Gy   Reference point between P-GW/PCEF and OCS to authorize usage of network resources in real time and report charging and resource usage information (e.g., used data volume in UL and DL). Gy is based on DIAMETER (see also Ro). For details, see 3GPP TS 32.251 [28].

Gz   Reference point between P-GW/PCEF and OFCS to provide charging relevant data (charging records) after usage of network resources is completed. For details, see 3GPP TS 32.295 [29].

ISC   Reference point between S-CSCF and AS based on SIP to provide services in IMS. For details, see 3GPP TS 23.228 [5].

M1   Reference point between MBMS GW and eNodeB for IP multicast delivery of user plane packets. The used protocol on M1 is GTPv1-U. For details, see 3GPP TS 36.300 [3] and 3GPP TS 36.445 [33].

M2   Reference point between MCE and eNodeB to convey radio configuration data for the multicell transmission mode eNodeBs and MBMS session control signaling. The used protocol on M2 is M2-AP (M2 Application Protocol). For details, see 3GPP TS 36.300 [3] and 3GPP TS 36.443 [31].

M3   Reference point between MME and MCE for MBMS session control signaling. The used protocol on M3 is M3-AP (M3 Application Protocol). For details, see 3GPP TS 36.300 [3] and 3GPP TS 36.444 [32].

MB2   Reference point between GCS AS and BM-SC based on DIAMETER. MB2 offers access to the MBMS bearer service. It has a control plane (MB2-C) and user plane (MB2-U) part. For details, see 3GPP TS 29.468 [26].

Mg   Reference point between MGCF and CSCF based on SIP to exchange session signaling messages for the interworking between IMS and circuit switched networks. For details, see 3GPP TS 23.228 [5].

Mw	Reference point between two CSCF (e.g., P-CSCF/I-CSCF and S-CSCF) to exchange SIP signaling messages. For details, see 3GPP TS 23.228 [5].
Mz	Reference point between BM-SC in Home Public Land Mobile Network (HPLMN) and BM-SC in VPLMN. It is based on DIAMETER. Mz is currently only supported for GPRS/UMTS, not for EPS. For details, see 3GPP TS 29.061 [10].
PC1	Reference point between ProSe application in the UE and in the server. It is used to define application level signaling requirements. This is not yet specified in 3GPP.
PC2	Reference point between ProSe AS and ProSe Function for EPC-level discovery. For details, see 3GPP TS 29.343 [23].
PC3	Reference point between UE and ProSe Function used to authorize ProSe Direct Discovery and EPC-level ProSe Discovery. For details, see 3GPP TS 24.334 [9].
PC4a	Reference point between HSS and ProSe Function. It is used to provide subscription information in order to authorize access for ProSe Direct Discovery and ProSe Direct Communication. For details, see TS 29.344 [24].
PC4b	Reference point between the SUPL Location Platform and the ProSe Function. For details, see OMA LIF MLP [34].
PC5	Reference point between ProSe-enabled UEs. It is used for control and user plane communication for ProSe Direct Discovery, ProSe Direct Communication, and ProSe UE-to-Network Relay. For details, see 3GPP TS 24.334 [9].
PC6	Reference point between ProSe Functions in different PLMNs or between the ProSe Function in the HPLMN and the ProSe Function in a Local PLMN. For details, see 3GPP TS 29.345 [25].
PC7	Reference point between the ProSe Function in the HPLMN and the ProSe Function in the VPLMN. It is used for HPLMN control of ProSe service authorization. For details, see 3GPP TS 29.345 [25].
Rf	Reference point for offline charging, for example, at the S-GW or BM-SC, to deliver charging events based on the DIAMETER Accounting application. For details, see 3GPP TS 32.240 [27].
Ro	Reference point for online charging, for example, between PCEF, TDF, ePDG, BM-SC, and OCS, to deliver charging events based on the DIAMETER Credit Control application. Ro includes functionality defined for Gy. For details, see 3GPP TS 32.240 [27].
Rx	The Rx reference point provides application layer information to the PCRF, for example, to establish new multimedia sessions and in turn appropriate EPS bearers. Rx is based on DIAMETER. For details, see 3GPP TS 29.214 [13].
S1-MME	Carries control plane messages between eNodeB and MME, for example, for bearer management, paging, and handover signaling; the used protocol on S1-MME, also called S1-C, is S1-AP (S1 Application Protocol). For details, see 3GPP TS 36.413 [30].
S1-U	Carries user plane packets between eNodeB and S-GW for the per bearer user plane tunneling; the used protocol on S1-U is GTPv1-U. For details, see 3GPP TS 29.281 [22].

S2a/b/c   These reference points are used to exchange control and user plane traffic when the UE is attached to a non-3GPP access system. They are based on PMIPv6 or GTPv2 (S2a and S2b) and DSMIPv6 (S2c). For details, see 3GPP TS 29.274 [20], 3GPP TS 29.275 [21], 3GPP TS 24.303 [7], 3GPP TS 24.304 [8].

S5   This reference point is used to tunnel user plane packets and manage the user plane tunnels between S-GW and P-GW. S5 is based on GTPv2-C or alternatively on PMIPv6. The huge majority of operators however have chosen GTP for S5, mainly as it is already used in 2G/3G PS domain and to avoid interoperability problems with other networks. For details, see 3GPP TS 29.274 [20] for GTP and 3GPP TS 29.275 [21] for PMIP.

S6a   Interface between MME and HSS to enable transfer of subscription and authentication data for authenticating and authorizing user access to EPS. It is based on DIAMETER. For details, see 3GPP TS 29.272 [18].

S9   Reference point between H-PCRF and V-PCRF to provide policy and QoS-related data from subscriber's home network to the visited network. It is based on DIAMETER. For details, see 3GPP TS 29.215 [14].

S10   Reference point between MMEs for MME relocation and information transfer, based on GTPv2-C. For details, see 3GPP TS 29.274 [20].

S11   Reference point between MME and S-GW, used to manage new or existing sessions, to relocate S-GW during handover, establish direct or indirect forwarding tunnels, and trigger paging. S11 is based on GTPv2-C. For details, see 3GPP TS 29.274 [20].

SBc   Reference point between CBC and MME for warning message delivery and control functions. The used protocol on this interface is the SBc Application Protocol (SBc-AP). For details, see 3GPP TS 29.168 [11].

SGi   This is the reference point between the P-GW and a PDN. Protocols on this interface are, for example, IPv4, IPv6, RADIUS, DIAMETER, and DHCP. For details, see 3GPP TS 29.061 [10].

SGi-mb   Reference point between BM-SC and MBMS GW for data delivery via IP unicast or IP multicast. For details, see 3GPP TS 29.061 [10].

SGmb   Reference point between BM-SC and MBMS GW for MBMS session and service area control. It is based on DIAMETER. For details, see 3GPP TS 29.061 [10].

Sm   Reference point between MBMS GW and MME for MBMS session control. It is based on GTPv2-C. For details, see 3GPP TS 29.274 [20].

Sp   Reference point between PCRF and SPR. Not standardized in 3GPP.

STa   This reference point connects the trusted non-3GPP access system with the 3GPP AAA server and transports access authentication, authorization, mobility parameters, and charging related information in a secure manner. The used protocol on this interface is DIAMETER (including DIAMETER EAP and NAS applications). For details, see 3GPP TS 29.273 [19].

SWa   This reference point connects the untrusted non-3GPP access system with the 3GPP AAA server and transports access authentication, authorization, and charging related information in a secure manner. The used protocol on this interface is DIAMETER (including DIAMETER EAP and NAS applications). For details, see 3GPP TS 29.273 [19].

SWm   This reference point is located between 3GPP AAA server and ePDG and is used for AAA signaling (transport of mobility parameters, tunnel authentication, and authorization data). This reference point also includes the MAG-AAA interface functionality. The used protocol on this interface is DIAMETER (including DIAMETER EAP and NAS applications). For details, see 3GPP TS 29.273 [19].

SWn   This is the reference point between the untrusted non-3GPP access system and the ePDG. Packets on this interface for a UE-initiated tunnel are routed toward the ePDG. This is a pure IP-based interface. For details, see 3GPP TS 29.273 [19].

SWu   This is the direct reference point between UE and ePDG to establish and maintain IPSec tunnels. The functionality of SWu includes UE-initiated tunnel establishment, user data packet transmission within the tunnel and tear down of the tunnel, and the support for fast update of IPSec tunnels during handover between two untrusted non-3GPP IP access systems. For details, see 3GPP TS 24.302 [6].

SWx   Reference point between AAA server and HSS to provide information about used PDN connections, APN, and AAA server address to the HSS. SWx is based on DIAMETER. For details, see 3GPP TS 29.273 [19].

Sy   Reference point between PCRF and OCS. For details, see 3GPP TS 29.219 [15].

References

[1] 3GPP TS 23.401: "GPRS enhancements for E-UTRAN access".
[2] 3GPP TS 24.301: "Non-Access-Stratum (NAS) Protocol for Evolved Packet System (EPS); Stage 3".
[3] 3GPP TS 36.300: "Evolved Universal Terrestrial Radio Access (E-UTRA) and Evolved Universal Terrestrial Radio Access (E-UTRAN); Overall Description".
[4] 3GPP TS 23.246: "Multimedia Broadcast/Multicast Service (MBMS); Architecture and Functional Description".
[5] 3GPP TS 23.228: " IP Multimedia Subsystem (IMS);".
[6] 3GPP TS 24.302: "Access to the 3GPP Evolved Packet Core (EPC) via non-3GPP access networks; Stage 3".
[7] 3GPP TS 24.303: "Mobility management based on Dual Stack Mobile IPv6; Stage 3".
[8] 3GPP TS 24.304: "Mobility management based on Mobile IPv4; User Equipment (UE) – foreign agent interface; Stage 3".
[9] 3GPP TS 24.334: "Proximity-services (Prose) User Equipment (UE) to Proximity-services (ProSe) Function aspects (PC3); Stage 3".
[10] 3GPP TS 29.061: "Interworking between the Public Land Mobile Network (PLMN) supporting packet based services and Packet Data Networks (PDN)".
[11] 3GPP TS 29.168: "Cell Broadcast Centre interfaces with the Evolved Packet Core; Stage 3".
[12] 3GPP TS 29.212: "Policy and charging control (PCC); Reference points".
[13] 3GPP TS 29.214: "Policy and charging control over Rx reference point".
[14] 3GPP TS 29.215: "Policy and charging control (PCC) over S9 reference point; Stage 3".
[15] 3GPP TS 29.219: "Policy and charging control: Spending limit reporting over Sy reference point".
[16] 3GPP TS 29.228: "IP Multimedia (IM) Subsystem Cx and Dx Interfaces; Signalling flows and message contents".
[17] 3GPP TS 29.229: "Cx and Dx interfaces based on the Diameter protocol; Protocol details".
[18] 3GPP TS 29.272: "Evolved Packet System (EPS); Mobility Management Entity (MME) and Serving GPRS Support Node (SGSN) related interfaces based on Diameter protocol".
[19] 3GPP TS 29.273: "Evolved Packet System (EPS); 3GPP EPS AAA interfaces".
[20] 3GPP TS 29.274: "Evolved General Packet Radio Service (GPRS) Tunnelling Protocol for Control plane (GTPv2-C)".

[21] 3GPP TS 29.275: "Proxy Mobile IPv6 (PMIPv6) based Mobility and Tunnelling protocols; Stage 3".

[22] 3GPP TS 29.281: "General Packet Radio System (GPRS) Tunnelling Protocol User Plane (GTPv1-U)".

[23] 3GPP TS 29.343: "Proximity-services (Prose) Function to Proximity-services (ProSe) Application Server aspects (PC2); Stage 3".

[24] 3GPP TS 29.344: "Proximity-services (Prose) Function to Home Subscriber Server (HSS) aspects (PC4a); Stage 3".

[25] 3GPP TS 29.345: "Inter-Proximity-services (Prose) Function signalling aspects (PC6/PC7); Stage 3".

[26] 3GPP TS 29.468: "Group Communication System Enablers for LTE (GCSE_LTE); MB2 Reference Point".

[27] 3GPP TS 32.240: "Charging Architecture and Principles".

[28] 3GPP TS 32.251: "Packet Switched (PS) domain charging".

[29] 3GPP TS 32.295: "Telecommunication management; Charging management; Charging Data Record (CDR) transfer".

[30] 3GPP TS 36.413: "S1 Application Protocol (S1AP)".

[31] 3GPP TS 36.443: "Evolved Universal Terrestrial Radio Access Network (E-UTRAN); M2 Application Protocol (M2AP)".

[32] 3GPP TS 36.444: "Evolved Universal Terrestrial Radio Access Network (E-UTRAN); M3 Application Protocol (M3AP)".

[33] 3GPP TS 36.445: "Evolved Universal Terrestrial Radio Access Network (E-UTRAN); M1 data transport".

[34] OMA LIF TS 101 v2.0.0, Mobile Location Protocol, draft v.2.0, Location Inter-operability Forum (LIF), 2001.

[35] 3GPP TS 23.468: "Group Communication System".

Index

access control, 183
   ACB skip, 185
   access class, 183–4
   access class assignment, 184
   Access Class Barring (ACB), 183–4
   special access classes, 184
   HighPriorityAccess, 183
Access Point Name (APN), 11, 12, 16, 35, 41, 47, 62
Access Stratum (AS), 14, 55–6
Accounting Data Forwarding (ADF), 140
Accounting Metrics Collection (AMC), 140
Address Resolution Protocol (ARP), 129
ADF, 142
All-IP network, 4
Allocation and Retention Priority (ARP), 17, 29, 53
allowed range, 134
ALUID, 134, 144
AMC, 140
announce, 112, 115
announcing UE, 110, 111
APCO, 7
APN Aggregate Maximum Bit Rate (APN-AMBR), 17
Application Code, 111, 142
Application Function (AF), 20, 36, 40
Application ID, 111, 141
Application Layer User ID (ALUID), 134, 144
ARIB, 1

ATIS, 1
Attach, 7, 18, 27, 197
Australian Communications and Media Authority (ACMA), 7, 9
Australian Public-Safety Communications Officials, 7
Authentication, Authorization and Accounting Function (AAA), 17, 45, 47

Base Station Controller (BSC), 6
Base Transmitter Station (BTS), 6
BCH, 153
billing, 34
Billing Domain (BD), 35
Broadcast Channel (BCH), 153
Broadcast Control Channel (BCCH), 29, 54, 171, 173
Broadcast Multicast Service Center (BM-SC), 38, 51, 52, 159, 163–7
BSC, 6
BSD, 153
BSR, 126
BTS, 6
Bucket Size Duration (BSD), 153
Buffer Status Reports (BSR), 126

Call Session Control Functions (CSCF), 39–40, *see also* CSCF
Capital Expenditure (CAPEX), 77
carrier aggregation (CA), 3, 191

LTE for Public Safety, First Edition. Rainer Liebhart, Devaki Chandramouli, Curt Wong and Jürgen Merkel.
© 2015 John Wiley & Sons, Ltd. Published 2015 by John Wiley & Sons, Ltd.

Printed and bound by CPI Group (UK) Ltd, Croydon, CR0 4YY

16/04/2025

14658381-0003